Love, Power
and **Meaning**

Love, Power
and Meaning

MYLES HOLLOWAY

DEIRDRE BYRNE

MICHAEL TITLESTAD

Additional material by

Verna Brown

Annamarie Carusi

OXFORD

OXFORD

UNIVERSITY PRESS

Great Clarendon Street, Oxford ox2 6DP

Oxford University Press is a department of the University of Oxford.
It furthers the University's objective of excellence in research, scholarship,
and education by publishing worldwide in

Oxford   New York

Auckland Bangkok Buenos Aires Cape Town Chennai
Dar es Salaam Delhi Hong Kong Istanbul Karachi Kolkata
Kuala Lumpur Madrid Melbourne Mexico City Mumbai Nairobi
São Paulo Shanghai Singapore Taipei Tokyo Toronto

and an associated company in Berlin

Oxford is a registered trade mark of Oxford University Press
in the UK and certain other countries

Published in South Africa
by Oxford University Press Southern Africa, Cape Town

Love, power and meaning
ISBN 0 19 571908 5

First published 2001
Reprinted 2002, 2003

Commissioning editor: Arthur Attwell
Editor: Martha Evans
Designer: Mark Standley
Cover Designer: Christopher Davis
Illustrators: André Plant and Bruce Titlestad
Indexer: Jeanne Cope

Published by Oxford University Press Southern Africa
PO Box 12119, N1 City, 7463, Cape Town, South Africa

Set in 10.5 pt on 14 pt Minion by Multi Media Solutions
Reproduction by Multi Media Solutions
Cover reproduction by The Image Bureau
Printed and bound by ABC Press

Contents

Introduction

WE BEGIN *Love, power and meaning* – a book about texts and interpretations – with a story. It is not one we have invented ourselves, for its roots lie in the origins of Western literature and, spanning thousands of years, it takes many unexpected twists and turns. In thinking about it, we will begin our exploration of factors central to interpretation and meaning.

The first part of our story comes from the Greek poet Homer's two epic poems – the *Iliad* and *Odyssey* – composed around 750 BC:

More than three thousand years ago, Helen, the beautiful wife of Menelaus, ruler of Sparta and brother to the king of Greece, Agamemnon, was seduced by Paris, one of the many sons of Priam, the king of Troy. Helen and Paris fled to Priam's walled city – Troy. Menelaus and Agamemnon were furious and sent a mighty army led by their bravest soldiers – Achilles, the two Ajaxes, Nestor and Odysseus – in a fleet of over a thousand ships to return Helen to her husband.

War followed. Locked in conflict for the next nine years, neither side made great gains. The Greek forces laid siege to Troy, but the Trojan resistance was strong. Finally, in the tenth year, Odysseus came up with the idea of hiding the best fighters in a huge wooden horse that was left outside the fortified gates of Troy.

The Trojans were astonished by this gigantic construction and furiously debated what they should do with it. Some argued that they should pierce it with their swords, some that it should be rolled off a high cliff, while others wanted it to be left as a signal to the gods. In the end, the argument of the last group won and the horse was wheeled into the city where it was soon surrounded by curious onlookers. Troy buzzed with excitement and wonder.

Suddenly, Greek soldiers – hidden in the belly of the horse – unlatched a secret door. In an instant, they were at the throats of their enemy, spreading fire and destruction. Blood, death, courage – heroic deeds of bravery and self-sacrifice – marked the battle. But eventually victory went to the Greeks. Troy was destroyed.

The second part of our story takes place in the 1870s:

For almost three millennia, Homer's tales were thought of as little more than the fantastical tales of a mythical past. Entrenched in Western consciousness as a result of the importance attached to classical Greek and Roman literature, the stories of ancient Troy – of Helen and Paris's doomed love, the weary years of war and the final fall of the city – fed the imagination of countless artists, writers and students, who read them, watched them enacted on stage or saw scenes depicted in visual art. But in 1870, a wealthy German merchant with a deep love of languages and ancient myths journeyed to the site of what he believed was the original location of Troy. The man's name was Heinrich Schliemann. A self-made millionaire and, by his own account, adventurer, devoted to the dream of finding the actual location of the Troy he had read about in Homer's epics, Schliemann set out to find the fact behind the fiction, the place about which epic tales of love, courage, tragedy and destruction had been woven for centuries.

Using geographical descriptions from the *Iliad* and following the advice of an American friend, Frank Calvert, Schliemann located a site near Hissarlik (in modern-day Turkey) that appeared to match the description of Homer's epic. In 1871, with a team of labourers, he sank a five-metre shaft, hoping to reach the oldest layer, which he believed would reveal the original Troy. He was to be disappointed. At various levels, his digging uncovered (and destroyed) evidence of earlier settlements, but failed to produce the dramatic find he wanted.

Schliemann wasn't deterred by this setback. The following year, with one hundred and fifty workers, he dug deep trenches all over the site. This time, he met with success. The remains of immense buildings and a substantial wall suggested that this was indeed the Troy that had so successfully kept the Greek army at bay.

Over the next twenty years, Schliemann funded and led many other more extensive excavations. They painstakingly peeled away layer after layer of earth and uncovered thousands of artefacts that spoke of an advanced civilization with a rich heritage of ritual, religion, industry and art. In his desire to prove the truth of Homer's writings, Schliemann identified many of these ruins and artefacts as objects named in the *Iliad*. His copy of the *Iliad* in hand, he declared one set of ruins the 'Palace of Priam' and another the 'Scaean gate'. He called a spectacular find of two hundred and fifty gold objects the 'Treasure of Priam'.

Yet, as the excavations continued Schliemann recognized that the site at Troy represented a series of civilizations, each built on the top of an earlier one. Not one city but many had existed in the place now known as Hissarlik. How did these finds relate to one another? What did they mean? What did they tell of life so long ago? Which one was Homer's Troy?

Archaeologists, then and now, have failed to reach a fixed conclusion. What is certain, though, is that the ruins that Schliemann so confidently claimed as proof of the Homeric tale predate the presumed siege of Troy by at least a thousand years. It is now believed that what Schliemann decided was Troy was (most likely) not the Troy Homer told of. Perhaps it was ancient Dardania, or forgotten, or another city lost in the mists of time, the traces of their histories virtually erased.

And now our story takes another twist.

While he was excavating Troy, Schliemann embarked on several other expeditions to locate and explore various sites mentioned in Homer's epics. His success in uncovering rich troves of treasure brought him great fame, but also many enemies. His critics have compared what he wrote about his methods and discoveries with eye-witness reports of his activities. Major discrepancies emerged, fuelling claims that he was less than accurate, open-minded or truthful. Schliemann had, it seemed, embellished the records to make his own activities seem more heroic and adventurous. The same critics also challenged Schliemann's interpretations of various objects, particularly when he claimed that they were the personal possessions of legendary figures like Agamemnon. Some critics went as far as to suggest that the artefacts were forgeries or that they had been gathered elsewhere and then hidden by Schliemann so that he could discover them later. Although many archaeologists believe that Schliemann's contribution to contemporary knowledge about Bronze-Age Greece is invaluable, the validity and worth of his project and its findings have been questioned.

We've begun this book with this story since we find at least two texts in it, each rich in ideas that are worth exploring. The first text is Homer's famous poetic account of the Trojan War. The relationship between the 'fictional' world of his poetry and events in ancient Troy raises interesting questions about art and life – about what we think of as imagination and what we think of as reality. As far as we know, such events did not really take place, although they may have had their basis in particular experiences. Moreover, although Homer is presumed to have composed these poems, almost nothing is known about him or the conditions in which they were produced. The Trojan War, if it occurred, took place at least five hundred years before Homer is thought to have lived. And the stories themselves were not invented by Homer; they were taken from an oral heritage that was deeply embedded in ancient Greece.

Yet, despite problems with the 'truthfulness' of the Homeric epics, Schliemann identified the site of ancient Troy (and many other places thought to be mythi-

THE SOUTH AFRICAN TROJAN HORSE MASSACRE

On 15 October 1985 in Athlone (Cape Town), ten policemen sat hidden in large wooden crates on the back of a truck. The country was aflame in the aftermath of the slaying of the Eastern Cape activist Matthew Goniwe, the declaration of a national state of emergency and the detention of the leaders of the United Democratic Front. When the truck carrying the crates encountered a group of protesters, the policemen emerged and opened fire using shotguns loaded with AAA buckshot and number one birdshot — both classified as deadly sharp-point ammunition. Many protesters were injured and three, Michael Miranda (11), Jonathan Claasen (21) and Shaun Magmoed (16), were killed.

At the time, the police claimed that this strategy was a 'necessary technique to protect the public from stone throwers' and, in 1997, Brigadier Chris Loedolff admitted to the Truth and Reconciliation Commission that the Athlone killings were not the only occasion on which 'this technique' was used. Eyewitnesses asserted, however, that the three people killed had not even been among a small crowd of protesters and that the police had fired indiscriminately into assembled groups of township residents.

This incident was filmed by a CBS television cameraman, Chris Everson, and caused an outcry when it was broadcast around the world. It became known as the 'Trojan Horse' shooting and was referred to as such throughout the Truth and Reconciliation hearings. Evidently, the 'Trojan Horse' has travelled a long way, and undergone many changes, from Hissarlik to Athlone. This shows that the easy distinction between stories and facts is seldom valid since we name, understand and judge the 'real world' through the stories that we tell. We often come to terms with our world by placing events, including those as traumatic as this, into stories and these will often be made up of the elements of other tales. In studying the stories of different cultures we are, it follows, engaging with more than incidental entertainment: we are learning about the ways in which people make sense of the world and their place in it.

cal), and contemporary scholars have shown that this city was in fact built and destroyed many times in its long history. Schliemann interpreted the 'story' of the siege of Troy in his own way and then looked for its 'truth', looking for some tangible proof of the myth. Put differently, he sought to unearth the fact in the fiction. As it turned out, he did not find the story he was looking for: he imposed a set of interpretations or supposed truths onto the site and elements of another story, another series of ancient cities and their cultures.

The plain at Hissarlik is another kind of text, though we might not easily recognize it as being textual. Hissarlik is a place of many lost civilizations and it also tells many stories, although the process of inferring and rebuilding these stories is extremely complex. Each trace of the past, such as objects found on the site, needs to be interpreted in its own right and against the evidence of all the other finds. This painstaking process requires much attention to detail as well as knowledge of other disciplines besides archaeology (literature, art, architecture and science, among others). Only after lengthy research may the archaeologist tentatively begin to tell the story of the site's past inhabitants – their behaviours and beliefs, their actions and their attitudes. Even then, the story that is told is a particular representation and reconstruction of the past based on selected and often incomplete information.

The plain at Hissarlik and the meanings that it can offer are very much like the texts that we encounter when we read poems and novels, watch plays and movies, listen to music or respond to artistic and commercial images. When we read these different kinds of texts, we unearth a complex web of signs that challenge us to give them meaning. This meaning is often elusive. In much the same way that Schliemann, in excavating Troy, named objects according to his own preconceptions and purposes, we as readers and viewers inevitably bring our own perspective, attitudes and values to the texts that we encounter. Depending on their soundness, these attitudes and assumptions have the potential to offer productive knowledge or to mislead us.

The things that we do as literary and cultural critics are similar to the things that archaeologists do. We respond to sign systems in texts as archaeologists respond to the material on site – we aim to unearth and identify those parts of the text that are most important for its structure and effects. For instance, in summarizing the ways that Troy has been 'read' and narrated in this book, we, too, have selected and interpreted what we think are the main points. Both archaeologist and critic are involved in processes of discovery and actions of interpretation. Both run the risk of misinterpreting the texts they read.

Love, power and meaning is primarily about interpretation. It deals with processes of discovery, of excavation and of ordering and evaluating information in the world of literary and cultural texts. The raw material of the archaeologist's attention are ruins and relics, while the literary and cultural critic is concerned mainly with the words, images and actions that make up the factual and fictional worlds that we encounter in books. Through them, we can enter into situations and circumstances that we have not personally experienced and worlds that no longer exist except in scattered ruins and ageing manuscripts. In the distant future, the books, painting and sculptures of our own society, along with the remains of buildings, household appliances and thousands of objects will be the signs the archaeologists will uncover, interpret and evaluate. What messages will they receive; what understanding will they reach?

Literary and cultural study, like archaeology, may seem irrelevant in today's world of big business and the global economy. The study of art, music or literature seldom brings instant wealth. Moreover, to preoccupy yourself with books or art at a time when violence, poverty and rampant crime mark daily life may seem self-indulgent, frivolous and blind to pressing social needs. Can we justify such study when the abuse of women and children, to name only one social horror, is routine. Are we wasting our time when gender inequalities and exploitation are the norm in most contexts? Why bother, when there are so many other urgent needs?

The conventional answer to this question is the claim that appreciating the arts, including literature, makes us better people. While the arts can obviously influence people, their overall effect is less certain. There is no inevitable connection between morality and writing or reading literature. Some of the most literate individuals have been responsible for the worst atrocities in history. Further, we cannot simply assume that individuals are personally enriched by the many texts that they encounter or that reading novels, plays and poems, or studying paintings and listening to music will lead them to a deeper understanding of themselves and their context. Obviously literature, art, songs and movies are about escape and repression as much as they help us understand who we are and the world in which we live.

It might seem surprising, then, that we do believe that the study of literature and other cultural expression is, beyond doubt, worthy of our attention. We live in a world that is increasingly, some would say entirely, formed by our diverse representations of it. Our very existence in the world is formed through the words (and, by extension, the phrases, sentences and stories) that we use to name and manage things, concepts and emotions. However, this language, these stories, are not only those that we read. Our idea of love, for instance, is conditioned by the representations of love that we encounter in television and radio programmes, song, religious teaching and in conversations with others. We cannot understand the networks of meaning of which we are a part unless we consider how representations emerge and how they compare with one another.

Roland Barthes, the famous French philosopher and literary critic, argued that, in spite of other influences, we cannot speak of love without 'quoting' literature. His argument is that the language in which love is depicted is deeply indebted to past literary representations. The very terms in which we express love are, according to Barthes, part of a history of love in which creative writers have been central. Literature and love, then, are closely related. It follows that we can come closer to understanding representations of love by considering its literary manifestations. Literature, if the example of love is anything to go by, can be considered as an important part of our systems of representation. Literary meanings both contribute to our representations and express meanings that circulate more generally in society. In summary, interpretation is an activity that allows us to reflect on the very ways that our existence and context become meaningful in the process of being represented.

Let's return for a moment to Troy, both as it is represented by Homer and encountered by Schliemann. Homer's and Schliemann's versions of Troy are specific representations of reality. Although Homer's creation is more obviously literary, Schliemann's text is not necessarily any more truthful. Neither text represents the first, the final or the only word on the subject. Homer harnessed an oral tradition to his own purposes, just as writers who followed him based their tales on The Iliad and The Odyssey. For instance, in the Aeneid (19 BC), the Roman poet Vergil takes up the story at the point where Odysseus' soldiers sack the city, but continues to follow the fate of Aeneas, who escaped to lead the other survivors to Italy. Many years later, Shakespeare returns to Troy in his play, Troilus and Cressida (1608) as have several writers since then. Similarly, Schliemann's reading of the artefacts excavated at Hissarlik has become just one of many, as his interpretations have been challenged, refuted or rewritten. What we understand Troy to mean today is a product of and a contribution to the meanings that Troy has had over the three thousand years in which it has been represented, created and recreated in the minds of men and women.

The way that the story of Troy has been re-interpreted and re-narrated over the centuries has an important bearing on the central themes of *Love, power and meaning*. First, it demonstrates that texts have an impact. For example, Homer could never have foreseen that his poetry would spark a series of archaeological investigations several millennia after his lifetime.

Second, what we call 'reality' (objects and experiences) interacts in complex ways with texts and representations. Texts shape the ways we think and, consequently, our identities. In turn, we create and recreate texts as readers, when we bring our experiences, values and assumptions to the process of interpretation. Finally, in focusing on these matters, this book asks you to learn to read with critical awareness of the dynamic forces at play in both reading and writing: to take responsibility for your critical stance and the interpretive choices you make.

 ## The Structure of the Book

Love, power and meaning is divided into three parts. Each part deals with a different subject or theme and is, in turn, divided into two chapters.

Part One, *Critical conditions*, introduces you to various approaches to literary study. It aims to allow you to develop a critical 'tool-kit' that you can take into the field with you as you explore various texts. More importantly, it explores the assumptions that underlie critical acts and encourages you to develop an awareness of the implications of treating texts in various ways. For example, the first chapter tracks the process of literary interpretation. There is nothing we can assume about the meaning of literary texts. The ways in which we deduce or infer their meanings are complex, and depend on what we assume about the relationship between authors and texts, as well as between texts and readers. Meaning is never independent of the processes through which it is uncovered or asserted. We thus have to consider the different ways in which texts can be interpreted, and decide which ways are most appropriate for particular works. In this chapter, you will debate questions such as whether it matters what an author 'meant' and whether or not the historical and political context of a text is relevant in interpreting its meanings.

As we interpret texts, we are inclined to judge them as good or bad, as successful or failing in some basic way. When you watch a film you probably grasp the 'meaning' immediately (which means you have interpreted it) and would often be inclined to tell someone that it is 'worth seeing' or not. This process of assessing the merits of a film or text is evaluation. The second chapter of *Love, power and meaning* looks at evaluation and asks on what basis we make decisions about the value of texts and how we can defend those decisions. In this chapter we will guide you towards forming judgements and expressing them

within reviews.

In Part Two, *Love, sex, body*, and Part Three, *Power*, we offer you an opportunity to exercise some of the critical insights that you will have gained from Part One. The texts that we focus on all require the types of reading and thought that we normally associate with literary and cultural study. But these parts serve another purpose that is closely aligned to Roland Barthes's claims about the importance of representation. Each part deals with a particular set of representations and ideas. They ask how these have come about and what their effect is. To some extent, we hope to trace three things: the origins of particular representations and the changing forms they take; the extent to which they shape our sense of what is worthwhile and valuable; and the extent to which representations reflect aspects of our own humanity.

For example, the chapters that make up Part Two focus on the body and on notions of love, sex and desire that are associated with bodies. The first one considers the ways in which bodies are used in advertising to sell commodities. We investigate how bodies are used in and even created by, advertising images. You are also encouraged to think about how women's and men's sexuality is manufactured in advertising through particular representational practices, and how these support (or challenge) patriarchal and heterosexual power. The next chapter picks up on these ideas, by considering different versions of and approaches to love and sex in poetry. While advertising and poetry might seem worlds apart, both rely on particular views of gender since they advocate specific ideas of maleness and femaleness. Further, especially if we concede that Barthes is correct that the language of love is always a quotation, then studying literary love means that we are also considering human emotional assumptions and practices.

In the final two chapters of this book, we look at varieties of political, social and personal power. If love is a primary site of literary production, so are power and politics. In fact, it is probably valid to assert that these two fields of human thought and activity have encouraged more literary production than any others. The first chapter in Part Three examines the idea of 'the political', demonstrating that politics is not limited to institutions or relations within and between nations, but that it pervades every aspect of the organization of our lives. Finally, in the last chapter of the book, we put this understanding of power into practice, by asking you to interpret *Dark Voices Ring*, a one-act play by Zakes Mda. We argue that every process of reading and writing is political. The very ways in which we see texts support particular sets of political assumptions. It is not simply that authors either are or are not 'political', but our decisions as interpreters and evaluators of literature are political decisions, which either support or challenge particular power arrangements

(what we might call 'configurations' of power). As soon as we pick up a pen as a literary critic (and you became a literary critic the moment you started reading *Love, power and meaning*), you are involved in political decision-making.

 ## Writers and Readers

Although all three authors collaborated closely when preparing the final manuscript of *Love, power and meaning*, each of us took primary responsibility for particular chapters. Michael Titlestad wrote the first three chapters, Deirdre Byrne the fourth, and Myles Holloway wrote Chapters five and six. In the process of writing this book we also had help and comments from colleagues, including Annamaria Carusi, Verna Brown, Gwen Kane and Karen Scherzinger, to whom we are indebted. Finally, we have had valuable assistance from Martha Evans, our editor at Oxford University Press.

Essentially, this book is the product of a co-operative effort – everyone has worked very hard to bring it to its final form. Co-operation does not mean, though, that we all agree about everything that has been written. The calmness of the final manuscript hides many heated arguments about the validity of claims and positions, and many disputes over what was worth including, when, why and how. These disputes are important because they point to the contentious nature of much of the material and from the fact that as individual academics, we have different opinions about literature, culture and criticism. Although we have tried not to contradict ourselves within the book as a whole, you will notice that the concerns and the emphases change as you move from one writer's chapter to another.

We hope that these changes in emphasis or differences in interpretive strategy will suggest to you that there is space for your ideas as a reader. We don't want you to 'learn' *Love, power and meaning*. We want you to respond to it and, if necessary, react against it with the full capacity of your intellect. This means that we don't require you to share our views or values. We are not advocating particular lifestyles or specific views on what is culturally and critically valuable. We don't expect you to agree with or like everything we say. Nor do we want you to think that we have a definite answer in mind lurking behind every question we ask. Rather, we would like you to see *Love, power and meaning* as opening up a new space in which you can think, read, write and make meaning. We hope that the space we have created is not threatening in any way. It certainly was not intended to be. Please excuse us if we appear to be bossy at times or if we ask you to answer questions that you find too personal or sensitive. Remember that you have the right not to read or to respond to anything that you find offensive.

Of course, this is not an invitation to skip any part of the book that you find difficult. The ideas that you will encounter are meant to challenge you into rethinking many aspects of art and life that we take for granted. Our arguments are often provocative or rely on subtle understandings of complex issues. To help you, though, we have included numerous exercises that will help you to focus your thinking and to sort out your own particular responses. Throughout the text, you will find information boxes (such as the one on the South African Trojan Horse Massacre) that either clarify points that we have made or provide additional, but related details. *Love, power and meaning* also has an index that will allow you to trace different places in the text where ideas are repeated. Finally, you will notice that we often invite you to talk to tutors, friends, fellow students and family members about the things that you are reading. These conversations, and the debates that may be provoked, form an important part of coming to grips with the ideas that are being introduced.

Apart from your careful attention, we also ask you to bring a few other things along with you as you read this text. First, we would like you to have a notebook or journal in which to record your observations and answers and in which you can record any questions that occur to you. Just as an archaeologist should keep scrupulous records of a dig, so too do critics need to keep a careful account of what they have read, what it means to them and how it relates to their experiences. Even more importantly, the texts that you encounter while reading through *Love, power and meaning* may inspire you to write creatively or critically. If this book does nothing more than inspire you to express yourself, it will have achieved much of its purpose. Secondly, we would like you to keep a dictionary at hand, preferably *The Oxford Advanced Learner's Dictionary*. This is your most basic guide to the signs that you will encounter most often – words. We would also encourage you to use other resources – libraries, encyclopaedias, the books that we refer to, critical sources, the Internet – to enrich your journey of discovery. Lastly, we would like you to bring along your own experiences, both in literature and in life. To get the most out of this book, you need to apply what you have read to other texts and to draw on your readings of other texts when assessing what we have written. More than this, you need to take the ideas that we have discussed out into the world. Make them your own. Let them inform your attitudes and decisions. Respond to them or reject them. Let them empower you.

Love, power and meaning suggests that nothing can be taken as given, that all aspects of our lives and those of our societies are made in language. To understand the language in which we and our world are constructed is to gain power. In entering the world of this text, then, while nothing is certain, this is also a place for you to make meaning, take a stand and seize the right to represent texts and ideas. We hope that you will leave this text with a sense of possibility; that you will believe that, in a reality made by the representations of others, we need to be critical, but also creative about the positions we take.

PART ONE

Critical conditions

OVERVIEW

In the first two chapters of *Love, power and meaning* we consider the interpretation and evaluation of literary texts. Chapter one deals with interpretation and Chapter two with evaluation. In these chapters we will guide you through some of the ways critics establish the meaning or meanings of literary texts. Every act of interpreting is also an act of positioning, or expressing your own views: the critic always writes from a particular perspective, as we will see. We will look at the (often unacknowledged) assumptions that are made when people interpret and evaluate texts. The chapters will also answer some of the questions you might have about the processes of reading, understanding and evaluating literature. Although the questions in Chapter one of *Critical conditions* focus on poetry and novels, they are relevant to plays, short stories and many other forms of cultural expression, such as films, songs and advertisements. As a result, this part functions as an important foundation for the rest of the book. The other parts of *Love, power and meaning* will refer to it often.

Work through these two chapters carefully, making notes and responding to the issues they raise. Try to come to terms with their concerns by thinking and writing about both the content and the implications of the discussions. As you work through the rest of the book, use the insights you have gained from *Critical conditions* to develop a more precise response to the texts and the issues we have chosen to highlight.

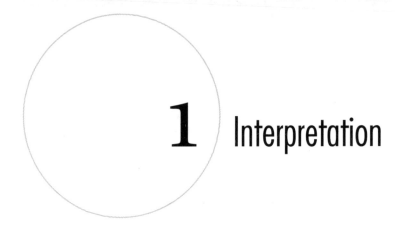

1 Interpretation

ONE

MEANING AND
COMMUNICATION

LET'S BEGIN THIS section by considering a particular exchange of communication:

You are driving along a dusty, deserted road, as the sun sets in the distance, when you notice a man standing at the roadside, waving his arms.

What is he doing? Perhaps he is doing roadside exercises to rid his body of tiredness from driving. Perhaps he is just

*being friendly towards passing motorists. Perhaps he wants
you to stop.*

✪ You then notice that he is standing behind a car, with
   steam pouring out of the engine.

*How does this help you to understand his gesture? How does
your understanding affect your next action?*

✪ Assume that you are a kindly, helpful person, and you
   pull over. The man approaches you and says: 'My car is
   overheating.'

*What is he implying? Is he telling you merely as a matter of
interest? Is he asking for help with his car? Is he asking for a
lift?*

✪ Assume that you are in a breakdown service vehicle.

*In what way does this help you to understand what he is
implying?*

✪ Assume that you are driving a taxi.

How does this fact affect your understanding of his utterance?

✪ Assume that you are a foreigner who does not speak English.

How much of the communication do you understand?

What emerges from this example is that in your efforts to understand the situation, you've had to *interpret* it. That is, you've had to grasp its *meaning*.

In trying to grasp the meaning of the situation, you've had to take various things into consideration: the man's gesture and statement, his circumstance (standing at a roadside, beside a car that is evidently broken down) and your own circumstance (the type of vehicle you are driving, and what this implies about the way you could help him). So your interpretation depends on the overall situation, which we refer to as the *context*. This shows that the meaning of a gesture or utterance is rooted in a particular communicative context, the particular situation in which the communication occurs.

Let's go over the process by which you came to understand the man's gesture and utterance.

Notice that we use the word 'utterance' in place of 'statement'. *The Oxford Advanced Learner's Dictionary* defines 'utterance' as follows:

▶ **utterance** /ˈʌtərəns/ *n* (*fml*) **1** [U] the action of expressing ideas, etc in words: *give* **utterance** *to one's feelings/thoughts/views.* **2** [C] words spoken; a statement: *private/public utterances.*

The word 'utterance' therefore covers all statements, including questions and commands, which the word 'statement' often excludes.

On first seeing the man waving, you were faced with a number of possibilities, ranging from stretching and friendliness to a request for you to stop. These possibilities lessened as you noticed a further feature of the context (the broken-down car), and it became evident that the last possibility was the most appropriate. This process was repeated as you came to understand his statement about the overheating of his car. Try to see for yourself how the context affected the way you interpreted his words.

But in order to complete the interpretive process, you and the man must share a common language, a common way of communicating. That is, you must both know the *conventions* of the language in which the utterance is spoken. This involves a shared vocabulary and a shared knowledge of the basic grammar of the language.

In any communication, such as the one sketched out above, or in any conversation, we are involved in a process of establishing the meaning of what is said, and this is the process of interpretation.

Communication consists of at least the following five elements:

```
        context          conventions
      speaker------utterance------hearer
```

The *speaker* produces the *utterance*, within a *context*.
The *hearer* interprets the *utterance*, within a *context*.

For example, consider the following: *Will my knocking wake you?*

While it's clear that this is a question from the order of words in the sentence and the punctuation (which are the conventions of the language), in order to begin interpreting it, we have to guess or deduce the following:

> ✪ What the speaker might have intended.
>
> ✪ The hearer to whom the utterance is directed.
>
> ✪ The context in which the utterance is spoken.

IN SUMMARY

Let's summarize what we have discussed so far:

✪ Interpretation is a way of understanding, which consists in grasping meaning.

✪ Every act of understanding meaning is an act of interpretation.

✪ To interpret an action or an utterance is to give it meaning.

✪ The meaning of a statement is related to the speaker's intention, which the speaker conveys by using the conventions of language.

✪ Interpretation also depends on the hearer sharing those conventions, and on the context in which the speaker and hearer find themselves.

· ·

INTERPRETATION

WE ARE INTERPRETING one another's actions and utterances all the time, but we don't usually think of ourselves as interpreting. We generally take interpretation for granted, and we use the word 'interpretation' only in special circumstances, like the following:

✪ When we are trying to understand the meaning of an utterance in a foreign language. In this case, interpretation means the same as *translation*. For example, think of the role of interpreters in court.

✪ In cases of misunderstanding, where a speaker might claim that he or she has been *misinterpreted*. Think about the following example:

Lebogang says to Mpho: 'You look nice, Mpho. Have you lost weight?' Mpho responds: 'Are you suggesting that I was fat?' Lebogang: 'No, you've misinterpreted me. I was just complimenting you.'

In this dialogue, misinterpretation is possible because the utterance itself is ambiguous. This means that the utterance could be interpreted in two different ways, either as a compliment or as an insult. The ambiguity might not be intended. In this example, it arises because the speaker's intentions (i.e., to compliment) are not clearly conveyed by the utterance ('Have you lost weight?').

So we tend to use the word 'interpretation' when we are consciously aware of the activity involved in understanding meaning – when we're interpreting foreign languages or ambiguous statements. This is the situation in which we find ourselves in our interactions with, for example, literary texts, films, plays, fine art and music. We need, then, to ask the question: What is it about these things that makes us so aware of the act of interpretation?

AMBIGUITY AND MISINTERPRETATION

In the course of World War II, a lookout on a battleship in the Irish Sea noticed a large metallic object bobbing in the sea about one kilometre off the starboard bow of the ship. Suspecting that it was a sea-mine planted by German saboteurs, the lookout ran below to the radio room. The radio operator sent out an urgent warning to all ships in the area: 'Unidentified object in Irish Sea, third sector. Think it's mine.' The only reply, received minutes later, was from a small fishing boat: 'Well, it certainly isn't mine.'

THREE

INTERPRETING

LITERATURE

NOW WE ARE going to look at what is involved in grasping the meaning of a poem. Poetic language is often more complex and formal than ordinary language. Exploring these complexities will help us to become more conscious of ourselves as active interpreters of literary texts.

The poem we have chosen is 'Biting Through' by Sue Clark.

BITING THROUGH
Sue Clark

Caught between doubt and desire
I fix on fantasy:
Imagine your close face
held between my hands.
Or a late night drive
narrow next to the sea
Then creeping beneath you
making fast to your sweet, safe mouth.

If I drive at night
will I find you home?
Will my knocking wake you?
Will you be alone?

Let me place
flowers in my room for you
Smooth cool pillows
Play sad songs –
Wait.

Suspended at this time of terror,
Keeping the glance of dark at bay.

(1996)

TIME TO WRITE

Answer the following questions in your journal. Keep your

answers. You will be referring to them regularly.

- ✪ Explain what you think the poem is about in three short sentences.
- ✪ Did you enjoy it? If so, why? If not, why not?
- ✪ Does the poem express an experience to which other people can relate, or is it only about the writer's feelings?
- ✪ What did you not understand about the poem? Which aspects did you find difficult?
- ✪ What are 'doubts' and what are 'desires'?
- ✪ What is the meaning of 'fantasy' in the second line of the poem?
- ✪ How do you understand the emotional conflict the speaker expresses in the first two lines?
- ✪ In what different ways does the speaker imagine the person for whom she is longing?
- ✪ What is the meaning of the three questions asked in the second stanza? Think of the doubts that the speaker in the poem mentions.
- ✪ Explain the importance of the actions set out in the third stanza.
- ✪ In light of your previous answers, suggest the nature of the 'terror' and of the 'dark' in the final two lines.
- ✪ Finally, what do you think the title of the poem means?

Now that we have read the poem and thought about its meaning, we are going to consider the question: what is it about this poem that makes us so aware of the act of interpretation? Here are some points you might have thought of while you were trying to grasp the meaning of 'Biting Through'.

Poetry is made of language, but language that is intentionally different from that which we use in ordinary communication. It is language 'made strange' through unusual combinations of words and images. Because it is 'made strange', our attention is drawn to the language itself, in contrast with ordinary communication where we are generally not aware of it. This, in turn, makes us aware of new ways of perceiving and experiencing things. Poets thus extend the range of language to achieve certain effects.

Think about the use of language in these two lines:

> Or a late night drive
> narrow next to the sea

The word 'narrow' is inserted into the sentence in an unusual way. The sentence is 'made strange' by this unexpected use of the word. This makes us question the meaning of 'narrow' in the context of the lines. What does 'narrow' refer to? How does it affect the way we think of the word 'drive'?

MAKING THE WORLD UNFAMILIAR

The idea that poetic language makes the world unfamiliar to us is a recurrent theme in literary criticism. Most of us live in a world of routine: we encounter similar things each day of our lives and represent them in speech that is largely consistent in its patterns. Perhaps because of these routines and habits of expression, we stop actively interpreting our world. When poetic language presents the world in language that is patterned differently, which links words and phrases that are not commonly combined, we are sometimes forced to reconsider the things described.

In 1840, Percy Bysshe Shelley, the Romantic poet, wrote in his famous essay 'A Defence of Poetry' that 'poetry lifts the veil from the hidden beauty of the world, and makes familiar objects to be as if they were not familiar' (cited in H. Bloom and L. Trilling (eds),

Romantic Poetry and Prose. 1973. London: Oxford University Press).

These ideas were interestingly developed by the Moscow Linguistic Circle. Viktor Shklovsky coined the term 'defamiliarization' to express the effect of poetic language. He argued that poetry specifically, and art generally, should

> impart the sensation of things as they are perceived, and not as they are known. The technique of art is to make objects 'unfamiliar', to make forms difficult, to increase the difficulty and length of perception, because the process of perception is an aesthetic end in itself and must be prolonged.

For Shklovsky, then, the role of art is to make us see the world anew, to prolong our perception so that we begin to see things in unusual, surprising ways. Art should, in the words of Shklovsky, 'make the stone more stony'.

The fact that we recognize a piece of writing as poetry leads us to respond to it in a particular way. As we read a poem, we begin to look for meaning, assuming almost automatically that there will be a complex relationship between the words on the page and what is being represented. We also notice a number of poetic devices, including the effect of connotation and the possible use of similes and metaphors. Although these devices are used in other texts, they are often seen as characteristic of poetry.

Sue Clark's poem has a number of possible interpretations. This is particularly clear in the last two lines and in the title. It's when we are faced with these different possibilities that we become aware of ourselves as interpreters.

Let's consider the last two lines of the poem:

> Suspended at this time of terror,
> Keeping the glance of dark at
> bay.

Here are some interpretations of the phrases 'time of terror' and 'glance of dark' in the context of the poem:

- ✪ 'time of terror' refers to the fear of crime occurring at this time in South African history.
- ✪ 'time of terror' refers to the speaker's fear of growing old alone.
- ✪ 'time of terror' refers to the fear of longings that will remain unfulfilled.
- ✪ 'glance of dark' refers to the speaker's fear of being observed by the person she loves.
- ✪ 'glance of dark' refers to the arrival of night across the sea (suggested by '... at bay').

POETIC DEVICES

It is worth looking for the following poetic devices when analysing poetry. They are often involved in the process of 'defamiliarization'.

- ✪ *connotation* the range of associations (other than simple dictionary meanings or denotations) that a word brings to mind. For example, in some contexts, the word 'red' doesn't only refer to colour; it also has connotations of passion, decadence and romance.
- ✪ *metaphor* a figure of speech that uses a word or phrase to describe something as another object in order to suggest that the two things share a common quality. For example, if you refer to someone as an angel, you are suggesting that he or she has angelic qualities. Metaphors can also be verbs (she blossomed) and adjectives (a green novice).
- ✪ *simile* an explicit comparison between two things, using the words 'like' or 'as', e.g., You are *like* an angel.
- ✪ *alliteration* the repetition of the initial consonants of words, e.g., tasty treats.
- ✪ *assonance* the repetition of identical or similar vowel sounds in the stressed syllables of words, e.g., hard heart, sweet dreams). It is not the same as a rhyme because the consonants differ.
- ✪ *enjambment* the running over of the sense and grammatical structure from one verse line to the next. For example:

  > Her heart skipped
  > a beat

  This is also referred to as a run-on line.

Arguably these are all ways in which these two phrases could be interpreted. Which do you think are acceptable interpretations? Why?

Different interpretations arise for various reasons:
✪ The 'strangeness' of the language creates ambiguities.
✪ Two elements of the communicative situation, the speaker and the context, are not readily accessible.
✪ In ordinary conversations, we constantly use clues – verbal, contextual and non-verbal – to help us understand our conversational partners. Even in these situations, we often fail to understand these clues.

How does this affect our interpretation of literature, where these clues are not available to us at all?

TIME TO DISCUSS

✪ Suggest two interpretations of the title of the poem: 'Biting Through'.
✪ Which of your interpretations is better? Can you choose between them?
✪ On what basis would you choose between your interpretations?

Choosing between interpretations makes us wonder why some seem more acceptable than others. Is it possible to offer an unacceptable interpretation of a poem? We need to decide on criteria for accepting an interpretation of a text. (A 'criterion' is a measure used to judge the value of something. The plural is 'criteria'.) Whether an interpretation is deemed valid or not depends on the criteria used. What one group of critics considers acceptable might be found unacceptable by another group. They do, however, have a shared understanding of the basic principles of interpretation.

The three basic principles of interpretation are coherence, inclusiveness and cogency.

Coherence: Here is the definition of 'coherent' in *The Oxford Advanced Learner's Dictionary*:

▶ **coherent** /kəʊˈhɪərənt/ *adj* **1** (of ideas, thoughts, speech, reasoning, etc) logical or consistent; easy to understand; clear: *a coherent analysis/argument/ description* ○ *The government lacks a coherent economic policy.* **2** (of a person) able to talk clearly: *He was not very coherent on the telephone.* **coherence** /-rəns/ *n* [U]: *narrative coherence* ○ *His arguments lack coherence.* **coherently** *adv*: *express one's ideas coherently.* Compare INCOHERENT.

QUESTIONS ABOUT INTERPRETATION

Here are some of the questions students ask about interpretation. Think of ways that you might answer them if you were asked to by a fellow student.

✪ Aren't all interpretations just a matter of opinion?

✪ How can you criticize my interpretation if it is my opinion?

✪ Is the lecturer going to mark my interpretation more strictly merely because she disagrees with it?

✪ Where do you get your interpretation from?

✪ Is it possible to learn how to interpret a literary text?

✪ How much do we need to know about writers when interpreting something they have written?

✪ Is the historical context in which a text was written relevant?

✪ Aren't all interpretations equally true/correct?

Coherence demands that your interpretations should not contain any contradictions. A contradiction occurs when two statements cannot both be true. For example:

> The table is round.
> The table is square.

Logic and the laws of language tell us that the same table cannot be round and square at the same time.

There is, however, an important difference between contradiction and ambiguity. Ambiguous statements can have more than one interpretation. These interpretations are not necessarily contradictory. For example, someone might say in a fabric shop: 'This fabric is lighter than I expected.' Here, the word 'lighter' is ambiguous since it can mean either lighter in weight or in colour. There is no contradiction involved in asserting that the fabric has both of these qualities.

Contradiction in an interpretation is unacceptable because contradictory claims cancel each other out. If you make contradictory claims you are, in effect, saying nothing at all.

Inclusiveness: Your interpretation should take account of as many relevant details of the text as possible. In principle, texts have many aspects (such as shape, typography, sentence structure, figurative language and so on) that could be relevant in an interpretation. Which aspects are relevant will depend on both the nature of the text and on the focus of the interpretation.

In 'Easter Wings', for example, the way in which the lines are printed is very important.

EASTER WINGS
George Herbert (1593–1633)

Lord, who createdst man in wealth and store,
Though foolishly he lost the same,
Decaying more and more,
Till he became
Most poore:
With thee
O let me rise
As larks, harmoniously,
And sing this day thy victories:
Then shall the fall further the flight in me.

My tender age in sorrow did beginne:
And still with sicknesse and shame
Thou did so punish sinne,
That I became
Most thinne.
With thee
Let me combine
And feel this day thy victorie:
For, if I imp my wing on thine,
Affliction shall advance the fight in me.

The poem is written in the Christian tradition. It is about the liberation of humanity through the death of Christ at Easter. The speaker claims that humankind will take flight if it 'imps' its 'wing' to Christ. The word 'imp' means to graft feathers on to a damaged wing so that it heals and the bird can fly again. Here, the suggestion is that if we connect ourselves to Christ, we will take flight into a greater freedom. Notice, too, the line 'Most thinne' occurs at that point at which the typography emphasizes the idea through the 'thinness' of the line. The line is mimetic of thinness. An interpretation of this poem that did not take into account its mimetic wing-shape would not be inclusive, as the shape reinforces the meaning of the poem.

But the typographic layout of a poem is not always as relevant as it is in this example. Furthermore, our particular interests and concerns will influence what we find relevant in texts. A feminist critic, for example, would be interested in matters of gender and power. Such a critic would look at the relations between the sexes, while another kind of critic might not. Take, for instance, Aristotle's definition of human beings:

Man is a rational animal.

If you assume that 'man' refers to all human beings, you would

POETRY AND MIMESIS

Poetry is an unusual kind of writing or speech in that a great deal of effort goes into *how* the poem is made, and not just *what* is said. To understand a poem, you must be able to read the *how* as well as the *what*.

Poets have at their disposal a wide range of poetic devices for making a poem. In poetry these are used in such a way as to 'act out' or imitate the point being made. You might find, for instance, in a line about a river, a rhythm that suggests a flowing motion. If that flowing rhythm is quick sounding, then you think of a fast-flowing river, but if the line's rhythm is slower, then you imagine a more sluggish passage of water.

Here's another example, from William Blake's 'London': the speaker of the poem notes that

...the hapless Soldier's sigh
Runs in blood down Palace walls.

Read this line aloud, paying attention to your body's own participation in making the meaning of the lines. As the first line ends, there is no comma to 'catch' your breath and so it escapes into the wide white space of the page, just like a sigh. Similarly, as your eye drops diagonally down to the next line, it 'acts out' the running action of the blood on the walls. In this way the poet reinforces the meaning of the content of the poem (*what* it's saying) with small acts of imitation in the form of the poem (*how* it is written).

Literary critics have noted how the form of poetry imitates its meaning since ancient times. The term by which the phenomenon is known is 'mimesis', which means 'imitation' in Greek.

GENDER AND LANGUAGE

It is worth thinking about gender and language in your own writing. It is easy enough to replace 'man' and 'mankind' with 'humanity' or 'humankind'. Similarly, instead of using the pronoun 'he' to refer to the individual in a generic sense, you can also use the formulation 'he/she'. ('The poet needs to make language strange: he/she needs to use language in an interesting way.') Alternatively, you can often replace singular pronouns with plural constructions. ('Poets need to make language strange: they need to use language in an interesting way.')

focus on the word 'rationality'. But if you were a feminist, the use of the word 'man' might be the focus of your attention. The word might be relevant to you in a way that it is not for other interpreters because it suggests that males are seen as the gender that defines the human race.

No single interpretation can take into account all aspects of a given text. But an acceptable interpretation must look at the important aspects of the text. However, what you might find important will depend on your position as a critic. You should include all that is relevant to your specific interpretive position. Different schools of criticism are discussed later in this chapter.

Cogency: *The Oxford Advanced Learner's Dictionary* defines 'cogent' as follows:

> **cogent** /ˈkəʊdʒənt/ *adj* (of arguments, reasons, etc) convincing; strong: *cogent evidence* ○ *He produced cogent reasons for the change of policy.* ▶ **cogency** /ˈkəʊdʒənsi/ *n* [U]. **cogently** *adv*: *Her case was cogently argued.*

A cogent interpretation is one that convinces the reader of its correctness. The critical reader must keep the same perspective throughout. If this is done, the different parts of the interpretation will be linked in a logical way. For example, when interpreting the poem 'Biting Through', the connection between the phrases 'time of terror' and 'glance of dark' should be looked at. A single interpretation would not interpret 'terror' as fear of crime and 'dark' as suggesting sexual desire. This reading would not be cogent because the two meanings are not compatible.

This does not mean that different perspectives cannot be adopted in different interpretations.

⃝ Inclusiveness, coherence and cogency: an exercise

Let's review these ideas by completing an exercise. Read the poem below and answer the questions that follow. The title of the poem and the name of the poet have been omitted.

> it is a dry white season
> dark leaves don't last, their brief lives dry out
> and with a broken heart they dive down gently headed
> for the earth,
> not even bleeding.
> it is a dry white season brother,
> only the trees know the pain as they still stand erect
> dry like steel, their branches dry like wire,
> indeed, it is a dry white season
> but seasons come to pass.

✪ How does placing the phrase 'for the earth' on a line on its own add to its effect?

✪ Would an inclusive interpretation have to take account of this? Why?

✪ If someone asserted that 'not even bleeding' suggests that the 'trees' do not suffer, would you agree? Explain.

- ✪ Would this interpretation be coherent and inclusive? Explain your answer.
- ✪ Does the 'whiteness' of the season indicate a snowy landscape? What evidence is there for your answer? If not, what does it refer to?
- ✪ The poem, written by the South African poet Mongane Wally Serote, is called 'For Don M. – Banned'. It was originally published in the volume *Tsetlo* in 1974. It is addressed to the poet and political activist Don Mattera, who was declared a banned person by the apartheid government in the early 1970s. How would this knowledge affect your interpretation?
- ✪ In the light of this, what would prevent a cogent interpretation being that 'broken heart' refers to an unhappy love relationship?
- ✪ What might the 'broken heart' refer to?
- ✪ In your opinion, do the final lines of the poem suggest political change? Write down your understanding of what they mean and how they work.

· ·

FOUR ○ Two interpretations of 'Biting Through'

IS ONE
INTERPRETATION
BETTER THAN
ANOTHER?

Let's see whether the three criteria of acceptability – coherence, inclusiveness and cogency – are adequate when choosing between interpretations. Read the following two interpretations of 'Biting Through', written by two different critics. Concentrate on both their differences and similarities.

INTERPRETATION A

THE OPENING LINES establish the emotional state of the speaker in the poem. She is 'caught between doubt and

INTERPRETATION B

THE FIRST TWO lines set the emotional tone of the poem. The speaker finds herself in a suspenseful and vulnerable desire', trapped between something she fears and something she wants. In this suspended state the speaker begins to fantasize about, or imagine, both her desires and her fears. She 'fixes' or focuses on 'fantasy' as a means of managing the sadness and loneliness she is experiencing. Her imagination, though, seems less a way out of her situation than a means of exploring it.

The three images the speaker imagines in the first stanza are moments of intimacy. The first image is of the face of somebody she loves (the beloved), the second is of a moment of intimacy on a late-night drive and the final image is of passionate love-making. The combination of touch, proximity and togetherness in these three images creates the impression of mutual love and devotion. These are the 'desires' mentioned in the first line. They may be desires that have been enacted or hopes for the future.

The third image is particularly interesting. '[C]reeping beneath' the beloved suggests both sensuality and security. The word 'fast' is ambiguous. *The Oxford Advanced Learner's Dictionary* defines 'fast' as: 'moving or able to move quickly; rapid' and 'firmly fixed or attached'. The reader does not have to choose one or the other reading. Both interpretations are valid. The speaker does not only rush to kiss the beloved in passion, but also attaches herself to the object of desire. The 'sweetness' and the 'safety' of the beloved's mouth suggest both the delight

situation. She would like to give herself over to desire, but her doubts (probably as to whether her desire will be reciprocated) hold her back. She is trying to find a way of living with the vulnerability to which her desire has opened her, but her means of doing this are avoidance and escape rather than facing up to her fears. This is suggested by the second line. She decides on fantasies as a way of keeping her problems at bay. Her imagining is a desperate act: she 'fixes on fantasy' in the way that a desperate person will clutch on to anything.

The rapid succession of these fantasies indicates the extent of her desperation. She grasps on to either memories or hopes of sexual and romantic intimacy, but can only keep them in mind fleetingly. She imagines her lover's presence, his 'close face' and her 'creeping beneath' him, but these positive thoughts are rapidly overwhelmed by the sheer negativity of the situation. Her doubts flood in, silencing her desires. She is obliterated in the space between what she wants and her fears.

The second stanza is not an explicit fantasy, as the first is. The language is far more ordinary. This is because the speaker has returned to the real world in which she is simply another woman who is subjected to male arrogance. The loneliness of her situation is emphasized by the fact that these (rather basic) questions would be answered in a relationship rooted in sharing and equality. The fact that the and the comfort gained from sexual intimacy.

In the second stanza, the images of desire give way to doubts. These are presented in language that is more direct because it contains fewer images. These are bald statements of fears. It is as if the speaker is brought down to earth, returning to the doubts that entrap her. Similarly, the language used is plainer and more down-to-earth. It is mimetic of the speaker's state of mind. The speaker's questions, and the uncertainty they express, indicate the emotional distance between her and the beloved.

In the third stanza, the speaker returns to her suspended state. She performs a series of small domestic rituals (placing flowers in her room, smoothing the pillows and listening to sad songs) to pass the time. These actions are an attempt to fill up the time of waiting, while preparing the house (and herself) for the possible arrival of the beloved.

The final two lines of the poem emphasize that the speaker is in 'a time of terror'. She is barely managing to avoid depression and loneliness. This is the 'terror' she confronts. She is, though, through her fantasies and actions, narrowly averting despondency. She is just managing to keep the 'glance of dark at bay'.

In the light of this reading, the title refers to endurance in the face of difficult circumstances. It alludes to the action of 'gritting one's teeth', an idiom that sug-

speaker cannot expect answers and that she is feeling pain shows that there is no honest sharing of emotions.

In the third stanza, we find the woman returning to the realm of fantasy. She prepares the room for a man who probably will not come. The flowers and the 'sad songs' are images of a funeral. We are witnessing the death of the woman's self as she desperately attempts to control the rising tide of despondency within her. Images of solitude and abandonment are everywhere: the pillows are 'cool' signifying his perpetual absence, she is 'suspended' and she is desperately trying to control the sadness ('the glance of dark') which might, at any moment, flood into her mind.

Keeping the images of suspension in mind, we can understand the title as signifying the 'biting through' of a thread from which the emotional life of the speaker is suspended. She is in danger because she has been left hanging in anticipation by the person she loves. To bite through would be to release herself, but also to fall into some frightening and unknown psychological place. 'Biting through' the thread of anticipation by which she hangs would end the suspense. The action would lead her to confront her fears, or to her fears being allayed. The fact is that by the end of the poem, these difficulties are not resolved. The speaker is still 'suspended', controlling her world through fantasy rather than through action.

gests the need to persevere no matter what the difficulties, the need to survive against all odds. The poem tells us about what is involved in continuing to love and desire someone in the face of doubts and fears.

One can see in the poem a common pattern of male power and women's failure to act. The poem represents a woman who is subjected to abuse by male power and arrogance. The person she loves forces her to wait while he fails endlessly to reassure her or engage with her on equal terms.

TIME TO WRITE

○ Are there significant differences between these interpretations? If so, what are they?
○ The first critic speaks of the 'possible arrival of the beloved', while the second states that the speaker 'prepares the room for a man who probably will not come'. Which of these statements seems more valid and why?
○ To what extent does the second interpretation give logical evidence for the statements 'she is simply another woman who is subjected to male arrogance' or 'there is no honest sharing of emotions'?
○ What do the two interpreters agree about?
○ Are both interpretations equally coherent, inclusive and cogent?
○ Which of the two interpretations do you find more convincing? Why?
○ Count all of the statements that seem to need further explanation. Can you find textual evidence for them?

We are now going to consider some points that might have made one of the interpretations more convincing to you than the other.

⊙ Factors that influence interpretation

Coherence, inclusiveness and cogency are the three basic requirements for any interpretation that we are likely to take seriously.
But there are other factors that help us decide if an interpretation
is justified. Here, we're going to mention three other factors.

> ✪ the context (both social and historical)
> ✪ the author's intentions
> ✪ the reader's subjective reaction

Read the following poem and decide what you think it might
mean. We will then look at three interpretations of it.

> THE SICK ROSE
> *William Blake* (1757–1827)
>
> O Rose thou art sick.
> The invisible worm,
> That flies in the night
> In the howling storm:
>
> Has found out thy bed
> Of crimson joy:
> And his dark secret love
> Does thy life destroy.

THE CONTEXT

Here is one interpretation of 'The Sick Rose':

> The sickness of the rose suggests the corruption of the body
> that occurs when it is infected with HIV. The rose becomes
> diseased and its life is destroyed when love infects its 'bed of
> crimson joy'. This obviously refers to healthy bodies before

they are infected with the disease that will eventually destroy them entirely. The love is 'dark' because it is evil and destructive, and 'secret' because obviously the man has hidden the fact that he is HIV positive from the woman with whom he has sex.

✪ Given the date when the poem was published (1794), could this be the meaning of the poem?

This interpretation applies an aspect of modern life (HIV) to a poem that was written long before the disease began to infect people. This is an example of an *anachronistic* interpretation. A meaning from a different historical period is imposed on a text. Interpretations like this raise questions about the importance of context (time and cultural situation). How important is the context of writing? Is it more important than the context in which the text is being read?

Some scholars, such as E.D. Hirsch, argue that one has to distinguish between two types of meaning in the context of literary interpretation: the meaning of the text itself, and the significance it might have for its readers. The meaning of the text, according to Hirsch, is determined

The Oxford Advanced Learner's Dictionary defines 'anachronism' like this:

anachronism /əˈnækrənɪzəm/ *n* **1** a person, a custom or an idea regarded as old-fashioned or no longer appropriate: *The monarchy is seen by some as an anachronism in present-day society.* **2(a)** the placing of sth in the wrong historical period: *It would be an anachronism to talk of Queen Victoria watching television.* **(b)** a thing placed wrongly in this way: *Modern dress is an anachronism in productions of Shakespeare's plays.* ▶ **anachronistic** /əˌnækrəˈnɪstɪk/ *adj.*

partly by the intentions of the author and partly by the conventions of the author's historical context. In the above example, critics who agree with Hirsch would argue that the meaning of the poem could not possibly relate to HIV since the disease had not emerged in the eighteenth century. Therefore, Blake could not have meant HIV when he used the word 'sickness'. In this interpretation, the text + the author = meaning.

However, the significance of the poem for the present-day reader could well include the suggestion of HIV.

Other scholars think differently: they feel that we cannot help interpreting texts from the perspective of our own socio-historical context. According to these scholars, critics are shaped by their own history and society and cannot fully understand another historical context. In other words, these scholars maintain that the original historical context cannot be recovered. We cannot travel into the past without taking along our knowledge of the present. Thus, our memory of the past is distorted.

In this view, the significance the text has for the reader (using Hirsch's terms) is its meaning: the text + the reader = meaning.

A major supporter of this view (that the past cannot be recovered) is Hans-Georg Gadamer. Gadamer's views of interpretation are based on the idea that the past and the present mutually influence one another. The present develops out of the past, and, at the same time, we only understand the past in relation to present circumstances. As a result, the past and the present cannot be simply separated.

In practice, we often interpret literary texts outside of the contexts in which they were written. In the process we create interpretations that are, to some extent inevitably, anachronistic. It is only when anachronisms are factually impossible, when they stretch that meaning beyond the limits of reason, that they cause serious damage to the interpretive process.

'Any interpretation of a historical text drags our present into the past.'

GADAMER

THE READER'S RESPONSES

Another interpretation might be:

When I first encountered this poem it expressed for me the difficulties of love. As soon as one establishes a relationship, the danger is that it will become its opposite. Somehow the joy of love is rapidly replaced by its corrupt and destructive counterpart. All love, in my experience, starts off as something beautiful and joyful, but rapidly becomes something that is possessive and destructive.

✪ How much does this interpretation actually have to do with the poem?
✪ How has the reader used the poem?

This is a *subjectivist* interpretation of the poem. The interpreter relates to the poem purely on the basis of his or her own experience and emotions. The interpreter seems to be more interested in his or her own response than in the poem itself. In critical discussion, we usually avoid subjectivism of this kind. Personal responses tend to be diverse and random, and do not help us arrive at a coherent, cogent or inclusive interpretation.

CRITICISM VS. ENJOYMENT

One aspect of the interpretive process you may feel we have neglected is enjoyment. People sometimes complain that analysis limits enjoyment, that looking closely at a literary work stops us from taking sheer pleasure in it. Pleasure in response to a text is subjective. However, it is also often the reason for much criticism. If there was no pleasure in the encounter with a text, why would we want to engage with it in the first instance? If we look at it like this, then criticism actually *is* enjoyment.

This doesn't mean that subjective views are completely irrelevant to the process of interpretation. Our first response to a literary text is often subjective. It is, however, important that the text limits our subjective responses. Simply asserting a subjective position does not give other readers a reason to accept your interpretation. It is only when subjective views are substantiated with reference to particular aspects of the text that they are useful. Subjectivism might be a departure point for your interpretation, but it is hopefully not the end point.

There is a school of thought, called reader-response criticism, which claims that a poem is in some sense a 'creation' of the individual reader. The reader 'makes' the meaning of the poem, and the poem exists as a meaningful entity only within the mind of the reader. This is a view put forward by Stanley Fish, among others. Fish gives the following amusing example, which actually happened:

Just prior to a group of literature students arriving in a lecture theatre, an English lecturer sees that there is a list of names on the board from a previous lecture. Instead of erasing the names, he decides to leave them on the board, and suggests to his students that they interpret what is on the board. Assuming that it is a poem, the students duly set about suggesting interpretations, and are dumbfounded when they are told that the list was not actually a poem.

✪ Could a list of names ever be a poem?
✪ Why do you think the students thought that the list of names was a poem?
✪ What ideas did they already have to have about poetry in order to interpret it?

Fish's point is that the list of names becomes a poem by the students' act of interpretation. Because they see it as a poem, it is one. A logical extension of this idea is that there are as many poems and as many interpretations of these poems as there are readers. If each reader can validly interpret the poem, then each might (hypothetically) create a different poem. However, since there is no shared object of interpretation, it would be impossible to have a meaningful discussion about the poem. In practice, we need to set limits to the ways in which texts are interpreted. These limits help us to convince others of the validity of our interpretations, particularly if they agree with the limits we have set.

The limits we set often give rise to interpretive communities. Once a given community accepts a set of procedures or limits for interpretation, they begin to have much in common with one another. They have a common basis on which to establish meanings and begin to share a language to express those meanings. So, if you were in a group of readers who all had the same views about literary meaning, that group would be an interpretive community. You would be likely to agree on your broad interpretations of texts.

THE AUTHOR'S INTENTIONS

Here is another interpretation of 'The Sick Rose':

> In this poem, Blake attempts to show the dark, corrupting nature of certain forms of love. He uses the 'worm' as a symbol of male love that is focused on possession and control and leads to corruption. Blake reflects men's love in this way because he wished to teach a different form of love to humankind – one rooted in the purity of the love of Christ uncorrupted by the way Christianity has redefined the figure of Jesus. We can only understand the meaning of the poem if we keep in mind the complex desire on Blake's part to give humankind a new understanding of love.

Some questions that arise from this interpretation are:

✪ How do you know whether Blake intended this?

Read 'The Chameleon Dance' composed by Michael Chapman. He took the wording directly from a newspaper report. Do you think it is a poem? Why?

THE CHAMELEON DANCE

The Minister of Home Affairs, Mr Stoffel Botha, has now disclosed that during 1985:

- 702 coloured people turned white;
- 19 whites became coloured;
- one Indian became white;
- three Chinese became white;
- 50 Indians became coloured;
- 43 coloured turned into Indians;
- 21 Indians became Malay;
- 30 Malay went Indian;
- 249 blacks became coloured;
- 20 coloureds became black;
- two blacks became 'other Asians';
- one black was classified Griqua;
- 11 coloureds became Chinese;
- three coloureds went Malay;
- one Chinese became coloured;
- eight Malays became coloured;
- three blacks were classed as Malay.

No blacks became white, and no whites became black.

✪ Do you need to know his intentions with certainty in order to interpret the poem? Explain.

✪ Would Blake's intended meaning exclude all other interpretations? Explain.

✪ How useful would it be if we could ask Blake to explain his poem to us?

This is an example of an *intentionalist* interpretation. There are a number of different forms of intentionalism. In its strongest form, intentionalism holds that the meaning of the text is identical with the author's intentions. More moderate versions claim that the intentions of the author are relevant to interpretation, but are not necessarily the key to the meaning of the text. Other variations see interpretation as a way of finding out what the author could have meant, given his or her socio-historical circumstances, and given the conventions of the language prevailing at the time. This form of intentionalism places the text firmly in the context in which it was produced.

There are, however, some problems with intentionalist interpretations. It is very difficult to know whether you have correctly grasped the author's intentions. Consider the example of rock art. In order to understand what the painters intended, one would have to understand a great deal about their society. Such reconstructions are very difficult because we do not have complete knowledge about these societies. So we can't know with certainty what the painters intended. However, historical and cultural distance is not the only barrier to grasping the author's intentions. Even the intentions of authors within our own time and society are very difficult to reconstruct. Authors are not always explicit about their intentions, and are often reluctant to talk about them.

The author's intentions are not the only factors deciding the meaning of the finished work. Intentions may change in the process of writing. For example, writers often speak of characters taking on a 'life of their own', so that the created character determines how the work proceeds, rather than the author's initial intention. In addition, the author's unconscious motivations may affect the process of creation. Literary works frequently have implications beyond the conscious intentions of their authors. Often, critics are able to point to aspects of the work that authors hadn't even considered.

SAN ROCK ART AND INTENTIONALIST INTERPRETATION

An important debate in prehistoric studies has centred on the interpretation of the rock paintings of the San people.

Two approaches to interpretation are discussed by Martin Hall in his work *The Changing Past: Farmers, Kings and Traders in Southern Africa*. The first is the common claim that the paintings should be seen as 'speaking for themselves'. According to this view, we should see the art as simply representing the world and life of the San. The paintings, then, are representations of the experience and life of a people who obviously had the desire to create and to leave traces of their world for later generations.

The second approach, on which most people now agree, is to apply the religious and social beliefs of the San to the paintings. One of the primary religious practices of the San was shamanism. This practice, which

occurs in many cultures, involves a medicine-man travelling to the realm of the spirits and in the course of his journey healing both individuals and the community. These practices mystically transformed the shaman into an animal while in a state of trance. If these mystical journeys are the subject of the many of the paintings, we cannot interpret them literally. Each animal and each human figure in the paintings has a deep religious significance and any meaningful interpretation would have to take account of this.

We find it difficult to interpret rock art because we do not know what the painters intended. Were they simply painting the world as they saw and experienced it, or was it their intention to reflect the complex religious processes of their spiritual lives?

Even if authors were explicit about their intentions, we would not necessarily be willing to replace the literary work with a brief statement on the part of the author. The work interests us, at least to some extent, in and for itself. Its ambiguities and complexities are appealing precisely because they create the spaces in which we ask interesting questions and come to our own conclusions.

Today, many critics question the idea that the author's intentions are especially relevant to critical interpretation.

Sue Clark

⭕ An author's interpretation: Sue Clark writes about 'Biting Through'

Let's look at what Sue Clark has to say about her own poem:

My feeling about a poem, or any work of art, is that it is the clearest way, at the time, for an artist to express intention. How intention is understood depends on the perspective and insight of the reader and on the quality of the work.

I fear that my explanation of 'Biting Through' is only one interpretation, but because I am the author it may reduce the involvement of readers who should bring their own imagination and experience to its understanding. Both interpretations that have been given are intelligent possibilities. I shall briefly give a third, focusing on some aspects of my intentions and understanding of the poem.

The first two lines and the final two refer to the broad context of the poem: 'Doubt' and 'desire' in South Africa in 1992 could refer to a number of issues, not all of them personal. The 'time of terror' was exactly that – there was a great desire for a just society after apartheid. Because of the efforts of elements such as the third force, there was a terrifying doubt that the desire would not be fulfilled.

The 'dark' that must be kept 'at bay' contains the terror of political uncertainty, but also covers other things that are frighteningly out of one's control: despair, loss, the inevitability of ageing, the 'glance' of death. These can be kept at bay or at least counter-balanced by intimacy, imagination, sensuality, irony, beauty, music, dreaming and poetry. The central section of the poem concentrates on these balances: lines 3–8 concern sensual yearning for someone living along the coast; someone recently met again after a number of years and desired again after a previously passionate but unsuccessful relationship.

Lines 9 to 12 describe the doubt of whether to act upon that desire in the light of previous experience with someone

CRITICAL CONDITIONS

who, although still unmarried, is unpredictable and secretive (line 10), gregarious and hard drinking (line 11) as well as sexually attractive (line 12). Lines 13 to 16 are displacement activities that can be trusted to distract and give pleasure while waiting, 'suspended', for some sort of certainty to emerge.

The title? At certain times of uncertainty in my life I obsessively consult a Chinese system of divination, called the *I Ching* or *Book of Changes*. 'Biting Through' is the name of one of the sixty-four hexagrams that may give one some indication of how to proceed with one's life. It seemed to be a suitable title for this poem with its sense of enduring as in 'biting the bullet'; also perhaps biting through the terror and taming it by turning it into a poem.

A POETRY PUBLISHER SPEAKS

Students of English don't often think about the intricacies of publishing. We tend to assume that excellent poetry will find its way into the world. We can only study 'Biting through' because it was included in the volume *The False Bay Cycle*, published by Snailpress. It is interesting to think about how fragile the process of poetry publication in South Africa is. Gus Ferguson, the owner and manager of Snailpress, was interviewed by Leon de Kock, the editor of the journal *scrutiny2*.

I know this is a hard question, but I feel compelled to ask it. You have said before that you set out, with Snailpress, to prove that it was possible to make the publishing of poetry in South Africa viable. How, then, are we to interpret Snailpress' closure, with a substantial overdraft, except as a sign that publishing poetry in this country is not viable. That it is, literally, a bankrupt affair?

It was my arrogant intention to prove that poetry publishing could stand on it own feet supported only by the poetry reading public. My plan was to produce slim, cheaply printed but beautiful books. The model was Sue Clark's *Winter Bounty* [her first volume] — my first production. It was nicely printed, well designed and had monoprints by Nicolaas Maritz. It sold out. It even went to a modest second printing. It made a profit. Everyone was paid and the poet received well-deserved royalties in advance.

Trouble is that my post-publication enthusiasm eroded my strict publishing boundaries. Snailpress and Firfield Press did cost me a lot of money. But 'bankrupt' is quite the wrong word, shame on you! May I quote Robert Graves: 'There might be no money in poetry but again there is not much poetry in money either.'

[Gus and Leon move on to discuss the effect of poetry reviewing in South Africa. You will be considering reviews in some detail in the next chapter, which concerns the evaluation of literature. Hostile reviewing, it must be remembered though, leads to fewer texts being bought and, consequently, limits publishing.]

Would you care to comment on the state of poetry reviewing in South Africa? What effect does it have on poetry publishing, in your experience?

Criticism in South Africa reflects the marginal position of poetry itself. Most reviewers are poets. Most negative or bad reviews seem to hold as paradigm for comparison with the reviewer-poet's own work. Caveat emptor. There is such a small readership that the reviews are read only (I suspect) by other poets. What upsets me though are not bad reviews of books but ad hominem reviews [in which the reviewer attacks the author and not the work] at the most banal level. Assumptions are made about the author's and publisher's intentions and there is often a most literal lack of imagination that assumes all poetry is directly biographical. There are of course good and generous critics. I have not changed my view of criticism because of my publishing experience. In a review I wrote on Mark Swift's *Seconds out* in *Contrast* some time ago I called for a more open reviewing policy to at least allow readers the chance to make up their own minds. Most poetry reviews in this country leave one indifferent to the book. Certainly critics have to be entertaining but not at the expense of the poetry.

*scrutiny*2 4(1) 1999: 32–3

. . . .

FIVE

QUESTIONS ABOUT INTERPRETATION

LET'S RETURN NOW to the questions asked on page 13. Given our discussion so far, we are in a better position to answer them.

Aren't all interpretations just a matter of opinion?
On one level, opinions are the origin of our interpretations, but they are not the end of the interpretive process. Convincing interpretations must include textual evidence to support what we think the text means. In setting out an interpretation, we express an opinion, but should not stop at the point of a claim. When we simply assert something and provide no argument for it, we give our readers no reason to agree. We should attempt to defend our claims by building arguments that are coherent, cogent and inclusive.

One of the exciting aspects of literature is that a text can make us change our minds. We might start out with a particular understanding of the text, and as we read or when we reread, the text might show us that we're wrong. For this reason, we have to keep an open mind and be flexible when interpreting texts.

How can someone else criticize my interpretation if it is my opinion?
Believing something does not make it valid or true. In all areas of interpretation (literary, scientific, philosophical, mathematical and psychological) we are expected to develop our arguments by defending our opinions and giving evidence for our claims. Anyone is (supposedly) free to hold any opinion. But 'opinion' and 'interpretation' do not mean the same thing – they are not synonymous. Opinions only become interpretations when they are expressed as arguments.

Is the lecturer going to mark my interpretation more strictly just because he or she disagrees with it?
Interpretations are judged on the basis of whether they are convincing or not. They should be written in such a way that you persuade your reader of their validity. If a person assessing your interpretation disagrees with it, it may mean that your argument is unconvincing: it may be illogical, it may lack evidence for its claims, it may contradict itself or it may leave out essential aspects of the text.

Where do critics get their interpretations from?
In the first instance, the details of the text determine what we can say about it. But we don't interpret these details in isolation. Rather, we do so within a context. This context includes the socio-historical circumstances in which the text was produced. It also includes the time and place from which we view the text: different societies and cultures will ask different questions of a text. We might also have particular concerns that frame our interpretations and the procedures we adopt in the process of analysis. For example, a feminist critic may focus specifically on the representation of gender and gender relations in a text. Finally, interpretations occur in different social situations; a discussion of a poem with a friend in a coffee shop is likely to be different from one written in an examination.

Is it possible to learn how to interpret literary texts?
Interpretation is an activity that is learnt. In a sense, interpretation is similar to chess: once you know the 'rules' there are an infinite number of ways to apply them and an infinite number of combinations that can be analysed and understood. Unfortunately, particularly for those learning to interpret, the 'rules' of interpretation (which are a complex series of practices and forms of thinking) are much more complex than the rules of chess. Interpretation is also a learning process: each act of interpreting teaches us more about the processes involved. Like most activities, we learn by practice.

How much do we need to know about writers when interpreting something they have written?
If your concern is mainly with the author (for example, if you are doing a study of the relation of an author's writing to his or her life), then biographical information is essential, as is information about the entire body of his or her work. One work can shed light on aspects of another, and such cross-seeding is an important aspect of the interpretation. This, however, is a specific type of study, and only one form of literary interpretation.

Generally, we should have some idea of the historical and social context of a writer in order to avoid anachronistic interpretation. However, any direct connection between biographical detail and the text is problematic. Think back to the issues we raised with respect to intentionalistic interpretation. Authors often try to distance themselves from their texts. There are, of course, some exceptions. Many authors write autobiographies or write autobiographical fiction or poetry (stories based on events in their lives). In these instances, you might well interpret a text with particular reference to the details of an author's life.

Is the historical context in which a text was written relevant?
Yes. Texts draw on a range of concepts and points of reference and these are specific to their historical contexts. Many texts, though, have some ability to speak across time. This is perhaps because they relate to aspects of reality that do not change as quickly as others. Some texts, however, have dated to the point of having very little bearing on the present.

CRITICAL CONDITIONS

Because texts are embedded within a socio-historical context, they provide valuable insight into the time in which they were written. They become useful resources about their historical periods. Many of our ideas of history and historical periods actually come from interpretations of literary texts.

Aren't all interpretations equally true or correct?

No. It is possible to misinterpret something in the text. It is also possible to construct an argument that is not valid (i.e., where the conclusions do not follow from the premises, or there is a mistake in one of the premises). If interpretations are described as true, we assume that they accurately describe a real state of affairs. For example, the statement 'South Africa is a country in Africa' is true because South Africa really is an African country. So if an interpretation is described as true, this means that the interpretation picks out features of the text and gives their meaning in a way that corresponds with the meaning of the text itself. The debate about whether acceptable interpretations are valid or true therefore hinges on the following question: does the text have an inherent (inbuilt or definite) meaning which interpreters uncover in their interpretations, and against which these

BAKHTIN AND HISTORICAL CONTEXT

Mikhail Bakhtin (1895–1975) was interested in literature and the voices marginalized by dominant groups. His concerns led him to initiate an important line of criticism which used the 'carnival' as its central image. During the days of carnival in many European cultures, all established order is turned on its head. Thieves are crowned king for the festivities and the 'grotesque' (i.e., absurd or distorted) aspects of life and society are celebrated. All authority is subverted and everything respectable is mocked. Bakhtin linked the practices of the carnival to literature's ability to subvert social power. He praised those writers and literary practices that challenged the order of things through celebrating the vulgar, the obscene and the playful.

The historical context of Bakhtin's work is central: he was writing under Soviet rule and, although his discussions seem concerned with distant history and writers from other cultures, his works are actually a celebration of the subversion of totalitarianism. His books were written in a mode that would not anger the Soviet censors. Historical circumstances were to impact on Bakhtin rather more directly, though. Although this story is probably apocryphal (it has been told often, but multiple embellishments make its truth questionable), it shows how fragile intellectual history might be.

During the Second World War, Bakhtin was supposedly trapped in Leningrad during a siege by the German army. In his possession was the sole manuscript of his most comprehensive work: a study of literary theory and philosophy. Bakhtin was in hiding with a group of intellectuals and friends. After a matter of weeks the group began to run out of food, but, more serious for the committed smoker Bakhtin, they ran out of cigarette paper. Faced with the dilemma of either preserving his supposedly great work of philosophy or being able to smoke — and provide his companions with that 'necessary pleasure' — he opted for a cigarette. More correctly, he opted for many cigarettes. By the time the group could leave their hiding place, the manuscript had gone up in smoke.

Not only does this anecdote illustrate history's impact on thinking, it also poses the interesting question: are ideas about art more important than life's necessary pleasures?

interpretations can be checked? Or is it rather the case that they don't have inherent meaning, but that interpretation is the reader's attributing or ascribing meaning to the text? In the first view, a literary text is like a box from which we pull things out (it is a process of discovery); in the second view, it is more like a box into which we put things (it is a process of creation).

So, whether we tend to describe interpretations as being valid or true depends very much on our understanding of what literature and interpretation are.

. . . .

SIX

CRITICAL POSITIONS (OR WHAT'S ALL THIS FUSS ABOUT LITERARY THEORY?)

HISTORY IS NEVER neat. Events do not naturally fall into the logical and clear stories we read in history books. We also seldom know, at the time of their happening, the importance events will have in the future. Written histories are a way of making sense of complex changes and any historical account has to leave out something (often a lot). Writing histories of real events in the world, like wars, political decisions or social changes, is difficult enough. Writing a history of ideas is almost impossible. Many people assume that, over time, humankind learns more, that we simply become more intelligent. What seems to happen, though, is that ideas change (almost like fashions) and often 'new' ideas are a matter of changing emphasis, of seeing other aspects of a process as important. What makes a history of ideas even more complicated is that people hold different views of things at the same time. At no point in time, for example, did everyone suddenly decide to believe that the earth was round rather than flat (some people, of course, still believe it is flat).

The history we map in this section suffers from both of these problems: it leaves a lot out and it creates the impression that ideas follow on from one another rather more clearly than they do. What we are going to look at is a history of the ways in which people have interpreted and evaluated literature. What we will see is that it is a history best understood as shifts in emphasis from one aspect of the process of literary communication to another. Let's begin by thinking through what this means.

We can understand literature as a process of communication. This may sound painfully obvious, but a writer writes a text to be read by an audience, and he or she does so in a specific social and historical context. There is a sender of a text (the writer, or in some communities, a storyteller or oral poet), the text itself (the book, poem, story, or spoken message), and a receiver of the text (the reader or, in the case of oral literature, the listener). Each of these three elements is located in time and place: writers belong to particular cultures at particular times, as do readers and critics. We have discussed at length the importance of context in interpretation, so this idea is by now very familiar to you. We could represent the process of literary communication and its context like this:

No critic discusses interpretation and evaluation as if only one of the elements matters. Obviously it would be rather odd to do so. It is possible, though, to emphasize one over the others, to see either the writer, the text, the receiver or the context as central to literary meaning. One critic might argue, for instance, that the author is the origin of meaning and the text merely a way of translating that meaning to a reader. Another might argue that meaning is created by the experience and knowledge of the reader when he or she encounters the words on the page (think back to Stanley Fish). Many critics believe that language is the source of meaning since no person can control all the complexities of language when writing, or decipher them when reading. For critics who believe this, the language of the text is primary (some even go as far as to say that we – our thoughts, emotions, even our subconscious – are 'made out' of language and we should therefore focus on it above all else).

Finally, there are some critics who argue that the relation of literature to its context is most important. Some say that literature reflects its world. For example, people's attitudes to race and gender will be mirrored in the text's written at the time, or that literature is used for various social and political ends. We will look at some of these positions, in rough historical order.

Like many historical processes, we cannot decide on the origin of 'literary studies'. However, ancient Greece seems to be a good place to start. One of the many things that the philosopher Plato wrote about was poets. He excluded them from his imagined utopian republic for fear that they might incite the population to violence by arousing their emotions. However, Aristotle (one of Plato's pupils) was more positive about art. He made a strong case for the ways that individuals benefited by watching, specifically, tragic drama. He claimed that by experiencing emotions while watching a play, destructive extremes could be controlled. Simply stated, if you empathize with the characters in a play and feel their terror or pity, these feelings are vented in a safe, controlled environment. While his view is concerned with society generally, he focuses on the individual's experience of art. This debate between the social and the personal function of literature is one that has raged continually over the ages.

It seems strange now, but for much of human history critics limited their attention to a study of Classics, to the literature of ancient Greece and Rome. They tended to remain concerned with the very works that Plato and Aristotle had in mind when they wrote. Although they read the literature of their day, critics felt that only the ancients were worthy of academic study. The writing of their age was, it seems, simply to be enjoyed rather than studied seriously. Most literary scholars agree that at the start of the twentieth century all this changed. Critics before then, among them the Romantic poets and Matthew Arnold, had made quite spectacular claims for literature. They believed that literature not only educated individuals about what was beautiful, noble and important, but also that this education would lead to a better society. However, it was only after the First World War (1914–1918) that these ideas were turned into an academic subject. Several modern critics, among them Frank Lentricchia and Terry Eagleton, have argued that the hold of Christianity declined rapidly after Europe's experience of the horrors of the war. A substitute had to be found, many believed, to teach the values that had been eroded by the decline of religion. Literary studies was in a good position to fill this gap. Not only does literature, according to the view of those who founded the discipline, express what is good and noble about humankind; it also teaches the values of culture, learning and beauty.

BIOGRAPHICAL CRITICISM

Interestingly, the first attempts at the formal teaching of literature seem rather unsystematic to us now. According to Terry Eagleton, they emphasized the author in the literary process. Much teaching of literary works amounted to telling stories about the lives of writers and looking for connections between their lives and the works they wrote. This form of criticism, which still exists today in various forms, is called biographical criticism. One of the assumptions of biographical criticism is that there is a direct connection between the author's life and the themes of a literary work. One might, for example, suggest that authors who write about happy childhoods do so either to reflect their own or to compensate for unhappy childhoods. Either way, the interpretation is affected by the author's biography. As you will remember from the earlier discussion of interpretation, we cannot assume a direct link between the author's intentions and what the work expresses. Similarly, it is difficult to assume a direct link between the lives of writers and the lives they create for characters. This does not mean that writers' lives do not matter when we interpret a literary work. Not only do people live in specific times, they also belong to specific cultures, have a particular gender and experience unique things. The danger is when we begin to make rather simple connections that do not take account of the fact that writers are creative, making worlds that may have little, in any direct sense, to do with their own.

RUSSIAN FORMALISM

'Art is a way of experiencing the artfulness of an object; the object is not important.'
  Viktor Shklovsky

If literary criticism was to fill the gap left by the decline of religion by replacing spiritual truth with secular truth, it would clearly have to be a more systematic procedure. Only then would it be something that could be taught so that it might have a sufficient impact on large numbers of people. In Russia in the early years of the twentieth century, a group of linguists (academics who study the features and workings of language) began to consider the characteristics of literary language. Literature, they insisted, was best studied by looking at what made its language different from other uses of language in society. In looking at the literary work, then, they separated it from its context, as well as from the author and the reader, and focused on how the text itself was constructed, on its formal aspects. For this group of critics, known as the Russian Formalists, the poetic devices in a text (for instance, the use of metaphors, similes, alliteration, rhyme, rhythm etc.) were its most important aspects.

NEW CRITICISM

'A poem has the structure of a fine and complex organism.'
  F.R. Leavis

For those in England and the United States who wanted to establish literary studies at university level, this approach was appealing. Since it established a definite object of study, the text itself, and proposed a definite system of interpretation, it seemed to offer possibilities for teaching while giving the process an almost scientific quality. It was as if the text was a specimen that one would dissect to find out how it worked. The school of criticism that developed is referred to as New Criticism, although there is nothing 'new' about it anymore. New Criticism is not the simple, misguided project it is often made out to be by many contemporary critics. New Critics argued many positions and formulated ideas that remain current in interpretation and evaluation. Among other ideas, they emphasized the connection between form and meaning, adopting in this a slightly moderated or 'softened' version of the Russian formalist stance. This connection suggests that words or phrases in a text should be interpreted by looking at both what they say and the formal devices used to say it, i.e., how they say it. New Critics also advanced the idea that the various parts of a text should be seen as inter-related, that the literary work is a unified whole in which every part is essential and dependent on the others. For this reason, one cannot interpret parts of a text in isolation from others. Since texts are complex combinations of various parts, the way to establish the meaning of a text is through close reading, a process in which you study every element of a text to see how it relates to the meaning and form of the others.

An aspect of New Criticism that many modern critics find problematic was their claim that literary texts expressed universal truths. If one is looking for secular truths rather than spiritual ones, this claim seems necessary. To assert, though, that literature expresses truths for all people for all time is problematic. The world, it can be argued, is a far more rapidly changing place than the New Critics ever imagined. In addition to change, there is no firm ground for claiming that all societies hold the same values or that they have the same ideas of, for instance, beauty and truth. Many critics in colonized countries have argued, with good reason, that the values of the New Critics were, rather than universal, the values of Western Europeans and Americans and that they serve their interests in terms of power and domination.

READER-RESPONSE CRITICISM

'Knowledge is made by people and not found.'
David Bleich

So far we have looked at three interpretive positions: biographical criticism, formalism and New Criticism. None has stressed the role of the reader in the process of creating meaning. Those critical positions that do stress the reader's importance are grouped under the broad category of reader-response criticism. Just as there are varieties of formalism, there are varieties of reader-response criticism. One version argues that every literary text is written with a reader in mind. The 'ideal' reader is, then, in the forefront of a writer's mind and this implied reader is the source of the text's meaning in an important and direct way. Another version, suggested by the American critic, Stanley Fish, states that readers produce the text through the way they interpret it. If you approach even a list of names on the board as a poem (remember the example in our earlier discussion), it becomes a poem. The author and, interestingly, the text are less important than the way the reader sets about interpreting them. Reader-response criticism has many contemporary versions, some of which look at the cultural and gender context of readers and their influence on how meaning is made and what evaluations of texts are developed.

MARXIST CRITICISM

'What is important about a work is what it does not say.'
Pierre Macherey

It was suggested at the outset that any literary communication occurs in a social, political and historical context. When interpreting a text one might emphasize neither the writer nor the reader, but see the text as reflecting something of the world in which it arises or as 'speaking to' the community that interprets it. Texts become, if one adopts a contextual position, a mirror of the society in which they were written or an engagement with the societies in which we live and read. Marxist critics see literature in relation to the economic organization of the society in which it is produced or read. Marxist theory asserts that all aspects of culture relate to the class structure and the division of labour at a given time in a society. Literature should, in classical Marxism, be read as reflecting (or at times opposing) the economic oppression of the working class. Early Marxist critics praised those novels that depicted the suffering of the working class and were openly critical of the industrialist and upper classes. Since then, though, variations of Marxist criticism have explored the relationship between language and power or have looked at the ways in which literature forms part of a complex web of ideas that has particular political and social consequences. Since Marx's theory was originally an account of the historical evolution of societies, all Marxist criticism considers literary works relative to their historical context, or looks at the uses to which they are put in different periods. Since Marxists believe that all values and ideas are historically specific, they oppose the idea of universal value advanced by the New Critics.

'ART exists in the real world and has a function in it.'

THEODOR ADORNO — *Marxist criticism*

HISTORICAL HERMENEUTICS

'No understanding without prejudice.'
Hans-Georg Gadamer

Marxist criticism is not the only theory that emphasizes the importance of the historical and social context of literature. Historical hermeneutics, which underlies many important contemporary ideas about literature, is based on the idea that we cannot

Hermeneutics although this term was first defined by religious scholars as the art and science of the interpretation of the Bible, the term now refers to any theory and practice of interpretation.

completely recover any historical context. According to historical hermeneutic critics, it is impossible to step fully outside of one's historical period into another. Any reading of a text from the past entails taking our world-view with us into the past. We cannot, for example, read a Shakespearean play without situating it, in some way, in the present. It follows from this that we endlessly reinvent the past in the light of the interests, knowledge and priorities of our own time. The complex connection between meaning and history is important. In the earlier parts of this chapter, we argued that it was necessary to know the historical context in which a text was written. While this remains valid in many ways, it is important to remember the limitations raised by historical hermeneutics.

POSTCOLONIAL CRITICISM

'Theoretical work must begin to formulate the relationship between empire and culture.'
Edward Said

Much modern literary scholarship adopts yet another version of historical contextualization, postcolonial criticism. Many critics argue that the literature (especially the 'oral literature,' or orature) of countries colonized by European powers has been neglected or interpreted solely in terms of the interests and priorities of European culture. They argue, not only that the literature of India, Africa, Australia, Canada and South America should receive more attention, but that the literature of colonizing countries (England, France, Spain, Portugal and so on) should be read in order to explore the extent to which it reflects or relies on the process of colonizing and exploiting other countries. This form of contextual criticism has a political aim: it seeks to redress an unequal balance of power that leads to the emphasis on European literature and the dominance of European standards and values. Postcolonial critics believe that, in spite of the apparent end of the colonial period, many inequalities relating to colonization continue in the modern world. What emerges from this brief discussion is that, while these critics do read literature in terms of history, they read it within or against a particular aspect of history, the European colonial domination of other countries.

FEMINIST CRITICISM

'The lost continent of the female tradition has arisen like Atlantis from the sea of English literature.'
  Elaine Showalter

Not all contextual theories are distinctly historical in their emphasis. Feminism concerns itself, mainly, with the ways that literature reflects or challenges the patriarchal domination of society and the social construction of masculinity and femininity as binary opposites. According to (most) feminists, society is structured around and by the interests of patriarchal power, which oppresses women. Feminists, like postcolonial critics, seek to correct a power imbalance in society. Initially many feminists sought to promote writing by women since it had been neglected, with few exceptions, by the (then) male-dominated universities and schools. Following progress in this regard, feminists turned to new ways of reading texts to reveal the ways that they reinforced or challenged patriarchy. Many contemporary feminist critics have shifted their focus to those aspects of writing and experience neglected in the traditional approaches to literature. Very important work on representations of bodies, on the significance of desire and on lesbianism has been undertaken in the last decade (as well as the construction of notions of masculinity and femininity). This work has also attracted the attention of scholars who have other interests in gender and sexuality, and new disciplines such as 'gender studies' and 'queer studies' have developed.

'Feminists have reacted bitterly to a view of women as passive, narcissistic, masochistic and penis-envying.'

MARY EAGLETON — *Feminist criticism*

POSTSTRUCTURALIST CRITICISM

'There is no outside the text.'
> *Jacques Derrida*

So far all of the schools of criticism (or critical positions) we have considered have emphasized one of the elements of the communicative process (writer, text or reader) or have looked at the context of the process as a whole. An intellectual revolution that began in France in the late 1960s was to change the way that people saw the processes of culture in general and of literature in particular. A perspective, known as poststructuralism, was initiated by a group of philosophers, the most famous of whom are Roland Barthes, Julia Kristeva, Jacques Lacan, Jacques Derrida and Michel Foucault. What united these philosophers, who worked in the fields of psychology, history, literary studies and the study of cultural change, was that they saw language as being formative. What this means is that language, rather than being merely a vehicle for communicating, is what 'makes' or constructs our world. According to the poststructuralists, everything is constructed by the way we name it and the ways that we use names in our stories about ourselves and our world. Our conscious and unconscious minds, our world, our histories and our values and morals are constructed by the words we use and the ways in which we use those words. There is, for the poststructuralist, nothing outside of language. This does not mean that the computer on which I am writing, the chair on which I am sitting, or the CD to which I am listening do not exist. It suggests, though, that the 'meaning' of these things, the way in which they occur to my mind and the significance they have for me, is rooted in the language in which they have been and can be represented. The point is more obvious if we look at an abstract example. 'Love', in a well-known essay by Roland Barthes, is described as simply a set of descriptions or representations in language. Our idea of 'love' is only a result of the way it has been represented (in language) over the centuries. What we think of as 'love' is therefore nothing more than a complex combination of

---

'Discourse is power.'

MICHEL FOUCAULT –
Poststructuralism

CRITICAL CONDITIONS

all of the utterances that have been made about love in the past. We are trapped in the present by historical shifts in meaning and any statement we make about the world relates, in the final analysis, to other statements rather than to anything we may have supposed to be real.

If we apply the poststructuralist view, not only texts but writers, readers and context are made of language. Reading is, then, the intersection of one complex text (the reader) with another (the literary work). Given that any utterance in a text relates to thousands of other utterances, there can be no final meaning of any part of a text and certainly not of a text as a whole. Biographical, New Critical, Marxist, feminist, or postcolonial interpretations often suggest a definite meaning of a work. Pure forms of poststructuralism argue that such interpretations deny the intricacy of texts, that they ignore the infinite complexities of reference that make up language.

An important aspect of poststructuralism concerns discourse. Discourse is, simply stated, language in use. Michel Foucault makes the point, in several of his works, that discourse is controlled and distributed in society along particular lines of power. An example might help to clarify this point. In modern times, people considered to be 'insane' are excluded from 'sane' society and are silenced through being institutionalized. However, there was a time when the speech of the 'insane' was considered to be prophetic or truth-telling. A specific discourse, the 'irrational' speech of 'insanity', has been excluded from the modern world. This exclusion has various effects. The most important is the support it lends the idea of rationality by establishing the discourse of 'reason' as the dominant (in fact, only) source of truth. From this example we can see that 'truth' is a function not of anything real, but of who we allow to speak and how seriously we take that speech. The distribution of the right to speak and to write creates knowledge in the world; and knowledge (if there is nothing outside language) is power.

Poststructuralism is a very complex field of philosophy, but its impact on studies of literature, history, art, psychology and culture has been substantial. While many people oppose poststructualism entirely, often claiming that it collapses important beliefs and values, it has altered forever the way we understand the relation of language to the world and the connection between power and knowledge. For this reason, it has had varied but profound effects on every other perspective and approach that we have considered in this chapter.

LIVES AND THOUGHTS

Just as we ask questions about the relation between writers' lives and their work, we can think about how theorists' lives affect their thinking and the reception of their philosophies. Here are just two interesting examples.

Paul de Man was a leading figure of the poststructuralist movement. In the words of the respected critic, Frank Kermode, he was 'the most celebrated member of the world's most celebrated literature school', namely the English Department of Yale University. Considered by his colleagues to be ethical and benign in spite of his formidable intellect, de Man was one of the few literary scholars whose claims about culture and aesthetics reached well beyond the confines of the academy. He was a spokesperson for his generation. It was, then, with amazement that the world discovered after de Man's death, through the work of Ortwin de Graf, a young Belgian scholar, that he had written for a pro-Nazi newspaper, *Le Soir*, during the Second World War. A student in Brussels at the time, de Man's writing for the newspaper was subtly anti-Semitic and could certainly be classified as collaborationist. This discovery raised important questions. How should this early writing of de Man affect our view of his later works and ideas? Should we now undervalue his contribution, especially in the field of the relation of ethics to art? There is no doubt that Paul de Man's status as a scholar was profoundly affected by Graf's discovery. Furthermore, in the wake of the scandal, the political implications of poststructuralism have been interrogated as never before.

Michel Foucault died of AIDS on 25 June 1984. There is little doubt that at the time of his death he was the most famous intellectual in the world. Foucault once argued that modern humanity understood itself mainly in terms of the ideas of Freud and Marx. Many scholars claim that we can now add Foucault's name to the other two.

Since Foucault's death, people have tried to reconcile his intellectual endeavours and his lifestyle. Just as Foucault wrote about the limits of our society, focusing on the exclusions we perform to create the myths of our 'normality', he also experimented with those limits through participation in an array of sexually extreme practices, including forms of sado-masochism. In the full knowledge of the emerging AIDS epidemic, he also took part in the promiscuous gay scene of the San Francisco bath-houses. At the same time, Foucault's writing turned away from the discourses of medicine, punishment and the social sciences towards the history of sexuality. There seems, then, to be an important link between his academic work and his life, perhaps united by his concern with 'limit experiences'. Sadly, it was his immersion in the limits of thinking and living that was to claim his life.

When I began this summary of critical approaches, I suggested that it would both simplify them and force them into a simple logic and sequence. In fact, many critical positions are carefully considered combinations or variations of those discussed here. Psychoanalytic criticism, for instance, looks at the text as a symptom of the unconscious mind of the author, at the process of identity formation in texts, or the use of desires in texts. It thus resembles biographical criticism in some variations, reader-response criticism in others and can also be a contextual position when looking at aspects of social psychology.

Similarly, queer theory, which sees sexuality as the organizing force of literary discourse, focuses on subversions of conventional representations of gender relations.

It remains to be said, though, that theory is not simply an open marketplace from which you can choose any combination of ideas. Any reading of a literary text has to be coherent, inclusive and cogent. You therefore have to choose a position and not change that position unless your shift is carefully motivated and the reasons for it made explicit.

2 Evaluation

WHAT DO YOU think of the following three works of art?

The Artist's Mother by Gerard Sekoto (oil on board canvas). Reproduced with permission from the Unisa Art Gallery.

(left) *Double Woman* by Walter Battiss (screenprint). Reproduced with permission from the Unisa Art Gallery.

(left) *Louis Maqubela* by Durant Sihlali (etching). Reproduced with permission from the Unisa Art Gallery.

> ✪ Which of the three paintings do you like the most and why?
> ✪ Which do you like the least and why?
> ✪ Do you feel that the Gerard Sekoto painting of his mother is a better painting than Walter Battiss's *Double Woman*?
> ✪ How do the Sekoto and Battiss paintings compare with Durant Sihlali's *Louis Maqubela*?
> ✪ Write a short paragraph to explain your preferences.

When looking at art, we tend to respond positively to some works and negatively to others. We often have quite a clear idea of which paintings we consider to be 'good art' and which we consider to be less good. Unless we know a lot about art, explaining our preferences might be difficult. In spite of this difficulty, we regularly make decisions about the relative merits of a work. When we decide whether a work is good or bad, we are engaging in evaluation.

Consider evaluation in our day-to-day lives.
✪ Do you regularly hear people evaluate films, songs and television programmes? What are the comments you most often hear?
✪ When you evaluate a film, song, television programme or book in casual conversation, do you simultaneously offer an interpretation of it?
✪ How are the ways in which we evaluate literature similar to or different from our responses to films, songs and television programmes?

INTERPRETATION, AS WE saw in the previous chapter, has to do with meaning. Evaluation, on the other hand, involves making a judgement about the value or worth of something. The process of literary criticism involves both interpretation and evaluation, and it's not always easy to distinguish between them. Still, it is important to be able to tell the difference.

MEANING AND VALUE

✪ 'Biting Through' is a poem that depicts loneliness and isolation.
✪ 'Biting Through' is a lovely poem.

The first statement is an interpretive claim because it tells us something about what the poem means; the second is an evaluative claim, because it judges the poem's worth. Evaluation is an assessment of a literary work according to particular criteria. If the work meets (or exceeds) the criteria by which it is judged, it will be positively evaluated. However, if the work fails to satisfy the critic's criteria, the critic will think that it has failed in some significant sense.

Since texts are written in language, any judgement of the value of a text involves first coming to grips with what the language means. If an evaluation of a literary text is to seem informed and convincing, it can occur only after we have decided what it means. When we give meaning to a text, we are often struck by its success or failure. As we read, we find things that we think are profound, interesting, moving, relevant, or the opposite of these positive evaluations. Evaluation, whether we are aware of it at the time or not, is a natural and inevitable part of the interpretive process.

Sometimes interpretive claims convey a value judgement. This is often true when such claims state the overall meaning of a text. For example, consider a statement like '"Biting Through" is a sensitive and profound expression of longing'. The words 'sensitive' and 'profound' are evaluative because they indicate the quality of the representation: they are measurements of the poem's success. But the evaluation is couched within an interpretive claim about the meaning of the poem – that the poem is an 'expression of longing' – which may be accurate. Interpretive and evaluative claims, then, are often combined or linked as parts of a single claim.

WHAT DO *YOU* THINK MAKES A GOOD WORK OF LITERATURE?

Fill in the following table to find out:

A good novel, poem, play:	Agree	Partly Agree	Disagree
is fun to read.			
uses complex language and symbolism.			
is morally true.			
is mentally challenging.			
influences people to change their lives and/or society.			
explores the intricacies of human existence.			
is highly imaginative.			
has a strongly developed plot.			
challenges conventional ways of thinking.			
promotes the personal development of the reader.			
is closely concerned with political and social issues.			
is elegantly written in grammatically correct English.			

CONSIDER THE FOLLOWING news headline, variations of which confronted readers throughout the world on 26 October 1999:

J.M. COETZEE'S *DISGRACE*: A CASE STUDY IN EVALUATION

J.M. Coetzee wins Booker for second time

J.M. Coetzee is one of South Africa's most respected novelists. He won the Booker Prize in 1983 for *Life and Times of Michael K.* and on the 25 October 1999 was awarded the prize (in absentia) for a second time for his new novel, *Disgrace*. He is the first author ever to be awarded the prize twice. The idea of a literary prize depends, of course, on a belief in evaluation. To choose which you consider the best novel from an extensive list is to apply a set of criteria to them and judge which best meets those criteria. We can learn much about literary evaluation by considering the Booker Prize, the controversies that surround it, and critics' opinions of the 1999 winner.

 ## The Booker Prize: a fact-sheet

The Booker Prize was first awarded in 1969, a year after its initiation. However, it only began to receive widespread attention in the early 1980s when it became highly lucrative and began to affect book sales significantly. In 1989 the prize money was increased to £20 000 (about R200 000).

In the United Kingdom and the Commonwealth it is the most highly regarded prize for literature. On a world-wide scale, it is perhaps second in status only to the Nobel Prize for Literature.

All books written in English by authors who are not citizens of the United States are eligible for selection.

Since its inception, the administrator of the Booker Prize has been Martyn Goff, who has achieved, because of this position, a unique status in the publishing and literary worlds.

Since 1977, five judges, including a Chair, have decided on the winner. Judges are chosen by the Booker Prize Management Committee, an institution riddled with politics and controversy. The ideal team comprises a respected academic, a literary editor of a quality newspaper, a writer, a qualified reviewer and a 'celebrity' to add 'the common touch' to the selection.

The selection process is also controversial. Publishers submit two titles to the committee for consideration. This is to control the number of titles the panel of judges has to read prior to compiling a shortlist. The long list is made up of between 100 and 130 books, which have to be whittled down to the short list of six in a matter of months. Some Booker prize judges have said that they had to read at least one book every two days for a three-month period.

It is reputed that hardback sales for the Booker Prize winner are enhanced by between 40 000 and 80 000 copies. Other shortlisted titles are guaranteed hardback sales of between 1 000 and 10 000.

The most commercially successful Booker winner to date is Roddy Doyle's *Paddy Clark Ha Ha Ha*. Having won the award in 1993, the novel had sold 360 000 in hardback and 340 000 in paperback by the end of 1994. The revenue these sales earned is probably in the region of £6m (about R60 million).

The least commercially successful Booker Prize winner was James Kelman's 1994 novel, *How Late It Was, How Late*. This was perhaps because it was also one of the most controversial, recounting, as it did, the consequences of a day of alcoholic blindness in a prose style littered with profanity. One critic pointed out that the word 'fuck' or one of its variants appears on average twelve times on each page of the novel.

It was in this competitive context that J.M. Coetzee's *Disgrace* emerged victorious in the Booker race of 1999. Before we consider critics' evaluations of the novel, it is useful to know something about its plot and concerns. Most book reviews include a description of the story prior to an assessment of its merits. Such a description, given that it must refer to the meanings of events in the novel, is also an interpretation. Even an apparently neutral retelling of the story is often a process of deciding on the meaning and worth of a novel. Keep this in mind as you read the plot synopsis below and the critical comments that follow.

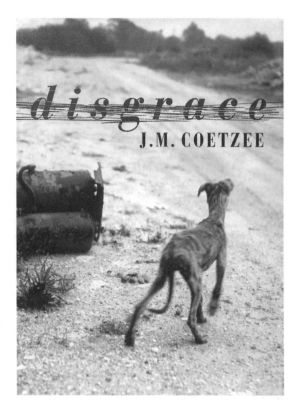

⬤ What is J.M. Coetzee's *Disgrace* about?

Disgrace recounts a series of events that occur in the lives of a 52-year-old lecturer, Professor David Lurie, and his daughter, Lucy, to whose farm he flees after a forced resignation from his job at the (fictitious) Technical University of Cape Town. Lurie, whose first love is the Romantic poets, teaches in an institution that has reduced English studies to 'Communications 101 and 201', two purely pragmatic language programmes. Following the collapse of a meticulously contrived and established relationship with an 'escort', Soraya, Lurie becomes 'involved' with a 20-year-old student, Melanie Isaacs. The last occasion on which the two have sex is, to all intents and purposes, rape, and the scandal that follows forces Lurie to resign. While he is prepared to admit his guilt to the University's disciplinary committee, his unwillingness to apologize angers them.

While Lurie is on his daughter's isolated smallholding in the Eastern Cape, events take an even darker turn. He and his fiercely independent and capable daughter are viciously attacked by three assailants. Lurie is doused with methylated spirits and set alight. He is locked in the bathroom while Lucy is gang-raped. Prior to the attack, Petrus, Lucy's neighbour and a landowner under the new post-apartheid dispensation, left the area without explanation. When he returns Lurie and Lucy discover that he is harbouring one of the attackers, a young mentally impaired boy. Lurie is deeply concerned for his daughter's well-being and

THE BOOKER PRIZE AND *DISGRACE*

Boyd Tonkin, the literary editor of the *Independent* and one of the judges of the Booker Prize, reflected on both the process of judging and on the winning novel.

Judging of the Booker Prize is often described as if it were a matter of five people sitting around a table and coming to the same conclusion simultaneously.

The reality, inevitably, is very different. Last night J.M. Coetzee's marvellous novel *Disgrace* won the Booker in its 30th anniversary year. His harrowing and profound fable of South Africa's predicament is as worthy and impressive a winner as the prize has ever had.

But to pretend this process was easy and without conflict would be ridiculous. As the chairman of the judges, Gerald Kaufman, said in his speech, this was a split jury. Anita Desai's account of the unhappiness of women in India and the United States, *Feasting, Fasting*, proved a serious competitor for Coetzee until the final stages of the judging. In a sense, it was a great privilege for the five judges to have to decide between two of the greatest writers of fiction in English at the present time.

Neither book makes for comfortable or optimistic reading. Yet both come from artists at the height of their powers who prove that the future of English literature depends, to a large extent, on writers whose origins lie thousands of miles away from the British Isles.

> FROM: Boyd Tonkin, 'Facts and fiction of judging the Booker Prize' in the *Independent* (*Enjoyment* section), 26 November 1999.

The fact that judges differed in their evaluation of the texts (or at least their opinion of which was the better) raises interesting questions.

- Is one critic's assessment of the merits of a text ever beyond dispute?
- Why, given the same text, might two critics' evaluations of it differ?
- Are all evaluations, then, simply a matter of opinion and could one argue that Wilbur Smith's latest novel, *Monsoon*, or an equivalent 'pot-boiler' is as 'good' as, or better than, *Disgrace*? The answer to this question is not, as many assume, self-evident or even obvious.

struggles to understand her choice neither to press charges nor to abort the baby she is carrying as a consequence of the rape. Lucy becomes increasingly dependent on Petrus and comes to consider an ongoing association with him as indispensable to her continued life on the farm.

The novel concludes with Lurie assisting at the local Animal Welfare Clinic, and Lucy as adamant as ever to find a way of belonging in this dislocated place and time. However, the final time that we see Lucy in the novel, she is pregnant, happily gardening and, according to her father, looking a picture of health. She seems, despite or through coming to terms with her trauma, to be more grounded than she was before and less despondent than her father, who resigns himself to the fact that 'one gets used to things getting harder; one ceases to be surprised that what used to be hard as hard can be grows harder yet' (1999: 219).

 ## What the critics had to say

We are now going to look at five different evaluations of *Disgrace* by professional book reviewers. They are presented here in extract, so appeared originally as part of longer, more systematic interpretations. Discuss the questions that follow each extract, keeping in mind your own criteria for the evaluation of texts that you established when you completed the table on page 54.

Paul Bailey, the novelist and critic, claims:

> *Disgrace* is a subtle, multi-layered story, as much concerned with politics as with the itch of male flesh. Coetzee's prose is chaste and lyrical without being self-conscious: it is a relief to encounter writing as quietly stylish as this. I was not totally convinced by Lurie's musical abilities, with regard to his proposed opera, but that is my sole complaint. The scenes at the animal clinic are wonderfully achieved, with Bev administering fatal drugs to the desperately sick and abandoned creatures, and Lurie taking their corpses to the incinerator at the local hospital.

FROM: Paul Bailey 'Beyond shame and tragedy: a damaged survivor edges towards hope in the new South Africa' in (electronic version) the *Independent* (Enjoyment section), 3 July 1999.

✪ If you favour a 'subtle' and 'multi-layered' story, what sort of story would you evaluate negatively? Can you think of an example of a novel that would fail to meet these criteria?

✪ What aspects of the novel does Bailey praise the most?

✪ Thinking back to the various critical positions discussed in the previous chapter, do you think Bailey's position is mainly a reader-response, formalist or Marxist one? Why?

✪ What, according to Bailey, is the novel's sole failure? In your opinion, is this a valid criterion on which to judge a novel?

The novelist, Barbara Trapido, born in South Africa but resident in the United Kingdom, argues that *Disgrace* reflects important aspects of the changing society it mirrors.

> With its pared-down, scalpel clarity and its mood of unrelieved sobriety, this novel says something close to the bone about the state of contemporary South Africa. It does so by choosing as its main character a representative of a dying class: a 53-year-old white male academic. This class is as surely being taken over by events as the Pasternaks and the Tolstoys of pre-revolutionary Russia. A tide is rolling against its liberal reformism, its high-minded, minority non-racism, and its traditional notions of academe. Its intellectually elite context is being 'transformed' out of existence.

FROM: Barbara Trapido 'Sensitivity training with dogs' in (electronic version) the *Independent* (*Enjoyment* section), 4 July 1999.

✪ Consider the first sentence of this extract. Is the sentence interpretive, evaluative or both? Explain how you came to your decision.

✪ How does Trapido's emphasis in her assessment and interpretation of the novel differ from Bailey's?

✪ Is she stressing the author, the text, the reader or the context of the novel in her evaluation of its strengths? What evidence would you put forward to defend your point of view?

On the day after the award of the Booker Prize, John Walsh discussed how a range of critics responded to *Disgrace*:

> Elizabeth Lowry wrote, in the *London Review of Books*: '*Disgrace* is the best novel Coetzee has written. It is a chilling, spare book, the work of a mature writer who has refined his textual obsessions to produce an exact, effective prose, and condensed his thematic concern with society into a deceptively simple story.'

> Boyd Tonkin, literary editor at the *Independent*, and one of the Booker judges, agreed. 'It's grim but fantastically well done. Every word is beautifully controlled, images from the early pages recur throughout the novel. It is a profoundly political book, but also a beautifully sustained prose poem.'

> > FROM: John Walsh 'A teller of spare and haunting fables that carry the weight of history. South Africa's leading novelist takes award for second time after jury splits' in (electronic version) the *Independent* (*Enjoyment* section), 26 October 1999.

✪ Both of the critics quoted by Walsh praise the novel by emphasizing its formal aspects (particularly its economical style and sustained imagery) and its social and political observations (its provocative reflection of post-apartheid South Africa). On the basis of this, would you say that evaluative criteria are completely arbitrary and can be combined at will? Why or why not?

✪ From what you know, or have deduced about *Disgrace*, why would a purely formalist evaluation be inappropriate?

✪ Is 'haunting', a term used in the title of this review, a positive or a negative evaluation in this context? Explain.

We conclude our discussion of the response to the novel with extracts from two South African reviews of *Disgrace*. First, consider Michael Morris's response.

DISGRACE AND RACISM

The controversial nature of Coetzee's *Disgrace* extended into the political realm. In the ANC's submission to the Human Rights Commission's hearings into racism and the media (in April 2000), a section from the novel was quoted as a proof of the racist mentality of white South African society. The ANC objected to the 'hordes of black savages as described by J.M. Coetzee' and argued that the novel shows that

> ... five years after our liberation, white South African society   •

continues to believe in a particular stereotype of the African, which defines the latter as immoral and amoral; savage; violent; disrespectful of private property; incapable of refinement through education; and driven by hereditary, dark, satanic impulses.

> CITED in Howard Barrell, 'A mean, insecure, fevered spirit abroad in the ANC' in the *Mail & Guardian*, 14 April 2000.

✪ What do you think of the ANC's evaluation of the novel?

It was once said of Coetzee that he had achieved a 'scrupulous disinterest' in the exploration of his world, and this is acutely true of *Disgrace*. Its great power wells, in part, from the steely dispassion with which events are shaped. At every turn, what is going to happen next is a pressingly awful question.

Hughes-Hallet says of the writer's approach: 'With merciless integrity, Coetzee withholds consolation, but he offers the reader the excitement of a grand project, achieved without a falter.'

> FROM: Michael Morris, 'Coetzee thinks publicly about new South Africa. New book describes society in a violent metamorphosis' in *Cape Argus*, 10 August 1999.

Now consider Winnie Graham's review:

Disgrace is a grim tale and yet, thanks to Coetzee's skilful writing, not entirely depressing. Indeed it has considerable relevance in a country where an atmosphere of uncertainty prevails. The tensions that go with major change and that are so prevalent in South African society engulf the reader

– yet life goes on.

What is it that makes the work so memorable? Undoubtedly it is Coetzee's remarkable restraint and understated dialogue. He tells it straight, exactly like it is. He does not overdramatise. The facts speak for themselves. More important, he does not lessen the impact by offering easy solutions. The South Africa he describes is a country its people will find all too familiar. This is Coetzee at his best.

> FROM: Winnie Graham, 'Tales of rape an allegory for South Africa' in the *Star*, 2 August 1999.

✪ Why might an author's ability for seeming 'scrupulous disinterest' be considered a merit? What negative effects might 'steely dispassion' have on a reader?

✪ The notion that a work should not be 'entirely depressing' is an interesting one. Do you agree that this is a valid measure of a novel's success?

✪ What other criteria do Graham and Morris see as relevant in their evaluations of the novel?

✪ The title of Graham's review, 'Tales of rape an allegory for South Africa', seems to support Bailey's claim that the novel is concerned, in part, 'with the itch of male flesh'. None of the reviews focused on the novel's representation of violence against women. What do you think a feminist reader would bring to his or her evaluation of the novel?

. . . .

YOU WILL RECALL from Chapter one that interpretations of a text can emphasize the formal aspects of a work, its context (the society and time in which it is written), the response of the reader, the writer's biography and intentions, or the gender, class or political situation of the author, reader or characters in the work. Similarly, if critics consider one or a combination of these areas to be important, they will accept or devise criteria that express this belief and use them as yardsticks to measure the relative success of the work. It is really because a reader interprets and evaluates simultaneously, using similar practices to determine

THREE

HOW IS
LITERATURE
EVALUATED?

meaning and establish merit, that the two are so often indistinguishable.

Let's consider some of the criteria for evaluation that emerged from the reviews of *Disgrace* and the discussions of its winning the Booker Prize. At least three distinct sets of evaluative criteria were invoked by the reviewers: *formalist, contextual* (social, historical and ethical) and *biographical*. We will discuss each of these in turn, and then discuss two other modes of evaluation that the reviews did not explicitly address: *feminist* and, what we might call, *popularist* evaluations.

 Formalist evaluation

A critic who adopts a formalist view of literature is concerned mainly with the words on the page and the writer's ability to use language in an interesting way. Formalist interpretations and evaluations focus on the technique and elements of composition of a work. They are typically preoccupied with plot structure, consistency of characterization, the compositional style, the quality of imagery, the use of figurative language, and so on. In its most extreme form, this type of evaluation focuses either on the success with which the writer has shaped the language or the way the various parts of the work harmonize with one another. The emphasis is not on what the work discusses, but rather on *how* its representations are crafted.

You will recall the praise of Coetzee's compositional style in the reviews of *Disgrace*. His novel was praised again and again for its sparse yet lyrical style, suggesting that he had achieved poetic use of language without needless flamboyance or ostentation. His capacity to write with an almost surgical clarity, yet highly evocatively, has been widely praised by reviewers and critics. You will remember, too, that his use of imagery was lauded by Boyd Tonkin, the Booker judge: 'images from the early pages recur throughout the novel'. Clearly, he considers (as do most critics) a consistency in intricately woven imagery to be a formal criterion of successful writing. This emphasis on imagery, like the emphasis on style, indicates a preference for formalist evaluation.

The danger of purely formalist evaluation is that the issues a work addresses may be ignored. If a finely crafted work expresses frightening ideas (say racist abuse), a formalist analysis might not engage with these atrocities. Formalism tends not to grapple with the moral and political implications of literary texts.

However, formalist criticism is rarely taken to this extreme. Most of the reviews of *Disgrace* combined a discussion of the style and imagery of the text with a concern for the changing world it reflected and the ethical implications of those changes. We need to remember, though, that one of the distinctive features of literature is that it is formally interesting and that it stretches the resources of language. It is logical, then, that this would be a yardstick. We cannot reduce literary works to a summary of what they are about, nor can we reduce them to simple mirrors of the time, place or political context in which they arise.

● IN WHAT WAYS do you take account of formalist aspects in your evaluation of what you see and read on a daily basis?

 ## Contextual interpretation

There are two aspects of the context in which a work is produced

---

DISGRACE AND SOUTH AFRICAN DISCOURSE: A CONTEXTUAL REVIEW

Disgrace stands in an interesting relation to the South African convention of the *plaasroman* (farm novel), as well as representing a new post-apartheid engagement with the issues of justice and racial (or cultural) difference. In terms of the *plaasroman* tradition, the novel inverts the Romantic idealism of nineteenth and early twentieth-century South African writing by representing the farm as a place of violence and contested ownership. In trying to achieve a sense of belonging, Lucy has to accommodate to a different identity and set of power relations that are manifestations of the new South Africa. Coetzee, in representing Lucy's dilemma, successfully engages with the *plaasroman* tradition, showing that surviving fragments of settler ideologies are not

enough to sustain either literature or white South African identities in our present.

Furthermore, *Disgrace* exalts its status as a post-apartheid novel freed from the contingencies of resistance and opposition. The social responsibility of South African writers has always had an impact on the mode in which ethical and political issues could be addressed. *Disgrace* is an important revision of the tradition of the South African novel in adopting a distinctly critical attitude to the dominant discourses of both reconciliation and apparent victory over injustice. Coetzee has introduced into South African fiction a new and important speaking position.

that may be seen as relevant to evaluation. These are the context of the literary tradition (for example, the tradition of the South African novel) to which the work belongs, and the social, historical and political context that the work depicts or in which it was written. Commentators who focus on the first type of context tend to see works as part of a particular tradition of writing and often ask whether the work is a fair representative or a significant revision of that tradition. They judge texts by comparing them to other works on the same subject or written in the same style. A danger of this method is that it can undervalue innovative works that break with tradition. New art is often seen as too 'way out' to be taken seriously.

Reviewers who focus on social context see works as reflecting or challenging a set of social realities or ideologies. Often they base their assessment on whether or not the work reflects the society adequately, accurately or insightfully. Often critics also believe that a work should fulfil a particular social role, such as political commitment or social responsibility.

Critics who feel that literature should have a progressive political agenda may feel justified in rejecting *Disgrace* on the basis that the emotional struggles of a middle-class white academic and his daughter are, in the final analysis, irrelevant to the important issues in our society. Such critics might argue that it is, at present, a South African artist's responsibility to support a programme of nation-building and an agenda of reconciliation rather than to ask profoundly challenging questions that urge readers to question the very basis of their optimistic ideologies. Questions concerning the political role of writing and critics' desires to

SOCIALIST REALISM AND POETRY

Think about poetry and politics as you read the extract that follows. It is a celebration of Joseph Stalin's 'supernatural' powers written by Krzysztof Gruszczynski (a Polish poet):

He who kept watch over the revolution like an eye,
He who led nations to victory,
He can direct the course of rivers,
He founds nature's laws.

It is Stalin who commands forests to march,
And orders rivers to flow into the deserts,
Stalin – our comrade, Stalin – our leader,
Stalin, the engineer of our dreams.

set the political agenda of creative artists are complex. Evidently, dictating a particular ideological line to authors is both unethical and artistically counter-productive. If we consider, for instance, the Socialist Realism movement in the Soviet Union, we can see how noble political aspirations can dampen artistic practices and cause the repetition of formulaic and uninspiring works of art.

Biographical evaluation

A biographical interpretation and evaluation looks at the text in relation to the writer's life. Biographical evaluation implies that there is a direct link between an author's experiences and their representation in a text. It assumes that sensitive writers will be able to write about their circumstances in a particularly vivid way, so that the reader can participate in and identify with the incidents that are depicted.

One of the problems encountered in biographical evaluations is that they tell us more about the author than the work he or she has produced. This often leads to a superficial interpretation of the text itself. Evaluations that linger over details of the lives and circumstances of authors tend to set them apart from ordinary people by suggesting that they are more sensitive or live more emotionally complex lives. Biographical interpretations tend to shift the focus of attention away from the work itself, towards the person of the author.

Biographical details cannot be dismissed out of hand, however. The author's situation, race, gender or political beliefs are often central to contemporary debates about literature. There have been various cases when a work has been presented under a false name and identity, and when this has been discovered, an uproar has ensued. When a work claims to be an authentic representation of actual human experience, biographical details are especially important. This might be particularly true when a work deals with the experience of an oppressed group (such as women, gays or particular cultural groups). Some critics would argue that only a member of the oppressed group that is being portrayed can write meaningfully about that group. A question that this raises is whether individuals from a particular cultural context can claim to speak as members of another culture. Generally, however, biographical criteria for evaluation play a relatively minor role in our view of a text's value.

- ✪ Why did the reviewers of *Disgrace* refer to Coetzee's identity and his past literary achievements in their evaluations of his novel?
- ✪ What links might there be between J.M. Coetzee's reputation and the way in which the manuscript and final text of *Disgrace* was treated by publishers and critics?
- ✪ How would the revelation that a text supposedly written by a black author and reflecting South African township life was in fact written by a white South African affect your evaluation of that text?

 ## Feminist evaluation

In our discussion of biographical evaluations, we suggested that the author's gender and experience of patriarchy might be pertinent to an evaluation of a text. The importance of gender in evaluative criticism does not stop there. A feminist critic might assess a work in terms of its recovery of the marginalized voice of women, by establishing the extent to which a text supports patriarchal power and reinforces stereotyped representations of women, or by measuring the work using the yardstick of the tradition of women's writing. This tradition has a different set of critical priorities and values aspects of literature significantly distinct from those favoured by patriarchal literary institutions.

Much current criticism is political in nature, assessing the merits of writing in terms of its political agenda or importance. Feminism is one such type of evaluation because it considers the gender context of the work to be central to any evaluation. This does not mean that all feminist critics undervalue formalist and other evaluative concerns. Their primary concern, though, is not the words on the page, but what those words tell us about the ways in which gender is constructed in the interests of particular versions of power and knowledge.

○ What systems of belief, political, ethical or religious, do you bring to bear on works when you evaluate them?
○ Do you think that a good work of art is one that promotes certain values and challenges particular beliefs or attitudes? Explain your answer.
○ Why do critics often judge art as bad when it promotes values with which they vehemently disagree?

Are there 'appropriate' criteria for evaluation?

There are many more approaches to evaluation than formalist, contextual, biographical and feminist. As is the case with interpretation, the approach used will depend on the critic's broader conceptualization and expectation of literature. Like interpretation, evaluation often varies from person to person. Whether the text is seen as interesting and valuable because of what it tells us about an author, because of its position in a literary tradition, because of its ability to reflect or grapple with social realities, or purely as a result of its formal features will be a matter of different and sometimes conflicting understandings of the nature of literature.

While we cannot argue that only one set of evaluative criteria can be a applied to a work, or that only a limited number of combinations is possible, certain sets of criteria do seem inappropriate to particular texts or literary traditions. It would not be inclusive to evaluate African American writing, for instance, without taking account of oral and vernacular (local) traditions in that culture or while denying the importance of the history of that tradition. It would also be limiting to conduct purely formalist analyses of politicized literature, or purely political evaluations of lyrical love poetry. A debate concerning the appropriateness of criteria of evaluation in South African literary criticism occurs later in this chapter. Many of these issues are discussed and developed there.

FOUR

A 'GOOD' READ: THINKING ABOUT POPULARISM

IMAGINE:

You have just finished reading a book that you thought was excellent in every respect. Because you enjoyed it so much, you recommend it warmly to a friend, looking forward to a pleasurable discussion of its many complexities and subtleties. But when your friend reads the book, she doesn't respond in the way you expected. She finds the book boring, flat and uninteresting.

This is a situation in which we have all found ourselves. Does it make any sense to ask whether either you or your friend is right? If evaluation is merely a matter of subjective opinion, it clearly doesn't. If it is only a matter of opinion, there is no reason to think that the poetry of William Blake, Ted Hughes, Seamus Heaney, Tom Paulin, Wole Soyinka or Mongane Serote is in any significant sense better than the verses inscribed in commercially produced greeting cards. In the opinion of some people the verses in greeting cards are profound and satisfying expressions of deeply held emotional truths. Others think that they are sentimental slush.

The poem that follows deals with this conflict in views: the popular versus the supposedly good. It is about Patience Strong, whose sentimental poetry is often used in greeting cards and popular magazines. Her work generally expresses positive sentiments, but in very simple and clichéd ways. In this poem, U. A. Fanthorpe, who worked for many years as a hospital administrator in the United Kingdom, uses the context of a hospital ward, and asks a very important question about literary evaluation.

PATIENCE STRONG
U.A. Fanthorpe

Everyone knows her name. Trite calendars
Of rose-nooked cottages or winding ways
Display her sentiments in homespun verse
Disguised as prose. She has her tiny niche
In women's magazines, too, tucked away
Among the recipes or near the end
Of some perennial serial. Her theme   perennial: perpetual
Always the same: rain falls in every life,
But rainbows, bluebirds, spring, babies or God
Lift up our hearts. No doubt such rubbish sells.
She must be feathering her inglenook.   inglenook: chimney corner
Genuine poets seldom coin this stuff,
Nor do they flaunt such aptly bogus names.   bogus: fictitious
Their message is oblique; it doesn't fit
A pocket diary's page; nor does it pay.

One day in epileptic outpatients,
A working man, a fellow in his fifties,
Was feeling bad. I brought a cup of tea.
He talked about his family and job:
His dad was in the Ambulance Brigade;
He hoped to join, but being epileptic,
They wouldn't have him. *Naturally*, he said,
With my disease, I'd be a handicap.
But I'd have liked to help. He sucked his tea,
Then from some special inner pocket brought
A booklet wrapped in cellophane,
Unwrapped it gently, opened at a page –

Characteristic cottage garden, seen
Through chintzy casement windows. Underneath   chintzy: of printed
Some cosy musing in the usual vein,   cotton cloth; also cheap and
And *See*, he said, *this is what keeps me going.*   mean

- ✪ What does the speaker feel about Patience Strong's poetry?
- ✪ Why do you think he or she feels this way?
- ✪ Why does the epileptic patient like her poetry?
- ✪ What does this disagreement tell us about the criteria for literary value?

RUSHDIE'S *SATANIC VERSES*

An extreme instance of different evaluations of a literary work dominated the news and debates about literature for nearly a decade. The Indian-born author and Booker Prize winner, Salman Rushdie, had a death sentence or 'fatwa' imposed on him by the Revolutionary Government of Iran because the content of his novel, *Satanic Verses*, was considered to be heretical. In their view, the novel not only lacked merit, it was a flagrant insult to the principles and people of Islam, particularly because of its representation of Mohammed. At the same time as the 'fatwa' was imposed (it was finally lifted in 1999), critics from various nations were heralding the novel as one of the most important works of contemporary literature. This difference demonstrates how divergent evaluations of a text can be if very different criteria are applied. We cannot debate fully the intricacies of the Rushdie case here, but it does show the potentially extreme results of certain forms of evaluation within particular political and religious contexts.

If we regard all evaluations as purely subjective, there is no principled way in which we can argue that greeting card poetry is better or worse than any other writing. Questions of value become impossible to resolve. However, if we shift our perspective away from individual subjectivity towards the judgements of a community or group, we may be able to argue that what is considered 'good' literature is relative to a social or cultural group, or a historical period.

People's tastes change over time. They also differ according to culture and socio-economic class. Expectations of 'good' literature, it can be argued, depend on the values and views of a society or culture. For example, deeply religious cultures are likely to have very different concepts of 'good' literature from secular cultures. This means that the value of a work depends more on the opinion of groups of readers than on its inherent characteristics.

The difference between this kind of cultural relativism and individual subjectivism is that it has added the important component of consensus or agreement. The unique Salman Rushdie case aside, judgements of texts based on agreement among people appear to be more objective than the opinion of one person and hence carry more weight. Obviously not all groups in society exert equal influence when it comes to evaluations of literature. One community that exerts an extraordinarily powerful influence on ideas about what constitutes good literature is the literary academy – scholars of literature operating from schools, colleges and universities. And it is this particular influence that we will consider in the next section.

. . . .

TRADITIONALLY, LITERARY CRITICS have argued that the works they consider to be 'good' are in fact good in some absolute sense. Underlying this assumption is the claim that the expertise of these critics in literary study qualifies them to distinguish *objectively* between good and bad literature. These critical judgements are often very influential. University literature departments promote particular types of writing (and, consequently, thinking) by prescribing the works of certain authors, while ignoring others. The works that are prescribed are taken seriously and acquire an aura of excellence and importance because of their selection. These selections follow patterns: the works of certain authors recur, the gender and racial bias of choices becomes apparent and particular modes of expression and complexity come to be valued over their contraries.

More obviously, though, these selections come to make up a list of 'great' literary works on the basis of shared evaluations by critics and academics. We refer to a list of 'great' works generated in this way as a 'canon'. Any selection of works that is promoted as 'good' literature or essential reading for an educated person is referred to as part of the canon.

In his book on the novel, *The Great Tradition* (1948), F.R. Leavis claims that the four great English novelists are Jane Austen, George Eliot, Henry James and Joseph Conrad. In his view their works are 'significant in terms of the human awareness they promote' and in terms of their 'genius'. According to Leavis, this 'awareness' elevates the works of the four novelists above those of other writers. Since Leavis, other critics, using similar criteria of evaluation, have made claims for the 'greatness' of a wide range of British and American writers in particular, arguing that they should be included in the canon. These claims have shaped English university studies worldwide.

FIVE

THE LITERARY ACADEMY AND THE CANON

F.R. LEAVIS

Frank Raymond Leavis (1895–1978) was an influential British literary critic whose teaching and writing profoundly affected the study of literature. Unlike other major critics of his time, such as the American New Critics, Leavis did not base his evaluation of literature on a close verbal analysis of the text. Instead he judged authors according to their capacity to experience life with maturity and sensitivity. His first book, *New Bearings in English Poetry* (1932) praised the work of T.S. Eliot, Ezra Pound and Gerard Manley Hopkins, but found the poems of W.H. Auden and W.B. Yeats immature. Later studies such as *Revaluation* (1936) and *The Great Tradition* (1948) contained similar lines of argument. Leavis's criticisms, which were often expressed with devastating scorn, found a wide audience through the journal *Scrutiny* (1932–53), which he edited with his wife, the scholar and critic, Q.D. Leavis.

A good example of the relationship between the literary academy and the influence of the canon is the treatment of the work of the metaphysical poet, John Donne (1572–1631). For many centuries, Donne was regarded as a bad poet. His poetry was regarded as too complex. It was felt that Donne's extensive use of paradox and of imagery drawn from science and alchemy made his poetry incoherent, even displeasing to the cultured mind. However, in the early 1920s Donne was elevated to the ranks of the great writers, largely because of the influence of the poet and critic T.S. Eliot. With this rediscovery came a resurgence of interest in Donne's life and work and in the publication of scholarly accounts of both. Today, Donne's poetry is studied by students of English across the world and his place on university and school syllabuses is well established. Few people recall his previous neglect or question his current value.

Of course, this does not mean that we should accept the literary canon as a natural or simple phenomenon. In fact, the existence and validity of the canon is a source of heated debate in contemporary literary studies. People who support it might claim that there are works that are clearly superior to others in terms of their formal aspects. They would argue that the canon is made up of works that can 'stand the test of time'. They can be appreciated by readers who live in very different historical circumstances from those for whom they were originally written because they convey 'universal' human values. Some critics also claim, like Leavis, that great literary works are morally uplifting or edifying and help to make the reader 'a better person'. Knowledge of the canon is seen as essential to a 'well-read' or 'cultured' reader.

The creation of the canon is open to attack. Opponents of the canon say that works do not have value across time and space. They point out that different societies at different times seldom agree on the value of literature. Similarly, there may be widely opposing views of literary value within a single context. Even Shakespeare, who is often regarded as the greatest writer of all time, has elicited contradictory responses. Further, there is no evidence to suggest literary works bring about moral improvement in their readers. As Terry Eagleton has pointed out, the commandants of the Nazi death camps during the Second World War were often well-read, educated people. Likewise, as we know from the South African apartheid experience, education and literacy do not guarantee moral or ethical behaviour. In addition, the canon is seen as limited because it privileges certain forms of literature. It does not allow for literary diversity. Autobiographies, transcripts of oral story-telling and works that use methods characteristic of the literary practice of other cultures tend to be neglected. The canon therefore offers a restricted view of what is accepted as valuable literature. Most seriously, perhaps, the canon of great English literature has been widely accused of being biased in favour of works by and about white, middle-class Western men. People from other groups, such as women, blacks, gays and people from non-Western cultures, have complained that the canon excludes them and positions them and their work as inferior.

One of the problems with the canon is that the criteria of selection are seldom made explicit. There seems to be a reliance on a silent shared understanding – virtually a 'gentlemen's agreement' – of what makes great literature. Even where elaborate claims are made for the merits of a particular writer or work, these may hide less explicit motives. In the case of John Donne's rise to fame, it was not simply because T.S. Eliot promoted his work that he came to occupy a position of prominence. Rather, Eliot's 're-discovery' of Donne occurred at a time when English was gaining popularity as a discipline and the New Critics (to

whom you were introduced in the previous chapter) were beginning to dominate the academic scene. In their attempt to show
that their subject was an objective discipline equivalent to any
other, the New Critics favoured complex texts that could be studied separately from their historical context or the reader's subjective response. Complexity was regarded as a merit because it
ensured that students and critics would always have something
to say. It allowed for debate and for the possibility of developing
finer and more detailed interpretations. Good students and good
critics were those who were able to penetrate the deepest, hidden
meanings of the text or those who could provide the most coherent explanations of texts that were elusive and contradictory. In
this climate, Donne was an ideal candidate for study. While critics praised his wit, the richness of his poetry or the inventiveness
of his paradoxical imagery, the unstated criterion of value was
his difficulty (which was previously the basis for his dismissal).

It is not surprising that canons and the claims that critics
make for 'great' literature give rise to feelings of alienation and
inferiority. This is especially true if one doesn't appreciate and
understand the 'great works'. Imagine being a young Indian or
African student in a classroom where William Wordsworth is
being described as 'great'. The way that Wordsworth's poetry
symbolizes early nineteenth-century English rural life may not
seem very important to you. As a result,
you may be categorized as barely literate
and uneducated. This is actually a common experience in colonial education,
where students are made to feel inferior
and their own culture is devalued. Some
of the reactions to this have been to
negate the Western canon completely
and to establish counter-canons.

DISCUSS THESE STATEMENTS:

✪ There are grounds for agreement
between cultures or at least for
cross-cultural dialogue about value.
✪ You can simply dismiss a canon on
the basis that it does not reflect your
own culture.
✪ Because the Western canon excluded
the literary expressions of other
cultures, one can dismiss it on the
grounds that it is white, patriarchal,
Western, bourgeois and so on.

AS BECAME CLEAR in our discussion of *Disgrace*, the evaluation of South African works of literature involves a complex set
of decisions concerning appropriate criteria. The controversies
surrounding J.M. Coetzee's novel are the latest contribution to a
long debate regarding South African literature. Thinking about
this debate helps us not only to contextualize the issues relating
to the evaluation of *Disgrace*, but to consider, at greater length
and in a specific context, many of the issues we have touched
upon so far in this chapter.

South Africa has a long literary tradition. Even before the colonization of the country by European powers, the indigenous
inhabitants enjoyed a rich and diverse oral culture in which
poetry, story-telling and drama were considered an important
part of social and religious life. With the arrival of the first set-
tlers came a written culture that included diary accounts, trave-
logues, letters, journalistic pieces and forms of writing that we
more readily associate with literary production – poetry,
fictional narratives and plays. Although these writings were primarily in English and Dutch, by the early nineteenth century
there were the beginnings of written literature in African languages. And by the start of the twentieth century, it is clear that
writing as an art form was well entrenched in South Africa. As
early as 1925, Manfred Nathan documented the works of four
hundred authors in his book *South African Literature*. When we
consider that Nathan's selection of South African literature is
mainly English and that his study ignores the work of black writers, it is clear that the actual extent of literary production in
South Africa was vast. Given its rich history, it is significant that
until recently our full literary heritage has been severely under-
estimated.

Before the early 1970s, the general view was that South African
literature was, at best, 'third-rate' and that South African writers
were not in the same class as British or American writers. School
and university syllabuses reflected this bias. Apart from isolated
works by a few writers, such as Herman Charles Bosman, Alan

SIX

SOUTH AFRICAN
LITERATURE:
THE EVALUATION
DEBATE

Paton, Athol Fugard and Guy Butler, students were not exposed to the vast majority of South African writers. Even when South African works were studied, the attitude of the literary academy was often negative. Critics pointed to lapses in style, to unevenness in characterization and to limitations in theme and content. For example, Olive Schreiner's *The Story of an African Farm* (1883), considered the first important South African novel, has faced severe attack. Some critics claimed that the plot is not developed, the characters are not fully drawn and the novel lacks organic unity. For instance, Kevin Magarey (an Australian) in an article titled 'The South African Novel and Race', claims that:

> ... *The Story* could without over-straining be described as a bad novel. It has powerful sections but an artistic incoherence that reminds one of George Eliot at her worst. The second, adult section seems to have little relation to, and is patently inferior to, the first, childhood one. The fiction is inadequate to embody the pilgrimage from Evangelical faith to pantheist rationalism; and the farm itself is inadequate, as setting or symbol, to unite the three roads into which the fiction branches in Part II. The book is thus forced to convey its message in the plain language of long dialogue digressions ...
>
> (FROM: *Southern Review*, p. 30, 1, 1963.)

Even local writer and critic Richard Rive says of Schreiner:

> Her plots are a series of episodes, and the reader's attention is distracted by long and endless dissertations on philosophy, art, evolution, allegories and whatever interests her ... the loose manner in which she puts down her thoughts, regardless of the principal theme of the book often detracts from the aesthetic value of her novels.
>
> (FROM: 'An Infinite Compassion', p. 34, *Contrast* 8(1), 1972.)

In essence what Magarey and Rive do is judge Schreiner's work according to the criteria used to interpret and judge works that comprise the canon of great English literature. They apply the criteria of formal realism, which favour a linear narrative, psychologically real characterization and plots that end with clear moral messages. It does not occur to either critic that the criteria they use to evaluate the novel might not be appropriate to it or to the experience the novel depicts. Schreiner's experiences in colonial South Africa during the nineteenth century were very different from those of, say, George Eliot or Thomas Hardy immersed in the heart of England. And, as Schreiner tells us in her introduction, she set out to 'paint the scenes among which [s]he has grown' (*The Story of an African Farm*, p. 24. 1985. Johannesburg: A.D. Donker).

Magarey's discussion of the South African novel goes on to raise one of the most common complaints against our literature – it is too preoccupied with political issues, particularly with questions of power and race. Ironically, the 1960s (when Magarey chose to criticize South African literature) and the 1970s saw a remarkable explosion of new writing by local authors. Much of this writing was by young black writers vigorously protesting the imposition of increasingly harsh racial legislation by successive regimes of the governing Nationalist Party. Art and literature were seen as key components in the struggle against apartheid.

FROM: 'GETTING OFF THE RIDE'
Mafika Gwala (1946–)

I ask again, what is Black?
Black is when you get off the ride.
Black is point of self realization
Black is point of new reason
Black is point of: NO NATIONAL DECEPTION!
Black is point of determined stand
Black is point TO BE or NOT TO BE for blacks
Black is point of RIGHT ON!
Black is energetic release from the shackles of
    Kaffir, Bantu, non-white.

...

MAFIKA GWALA was a pioneer of Black Consciousness. Apart from two volumes of poetry – *Jol'iinkomo* (1977) and *No More Lullabies* (1988) – Gwala was a leading spokesperson for black South African writing during the 1970s. He has also published several critical articles and short stories. (Adapted from David Adey *et al*, *Companion to South African English Literature*, Johannesburg: A.D. Donker, 1986)

I hear the sound of African drums beating
to freedom songs;
And the sound of the Voice come:
     Khunga, Khunga!
  Untshu, Untshu!
       Funtu, Funtu!
  Shundu, Shundu!!
  Sinki, Sinki!
          Mojo, Mojo!
O-m! O--0--m! O---hhhhhhhhhmmmm!!!
The voice speaks:
'I'm the Voice that moves with the Black Thunder
I'm the Wrath of the Movement
I strike swift and sure
I shout in the West and come from the East
I fight running battles with enemy gods
  in the black clouds
I'm the watersnake amongst watersnakes
  and fish amongst fish
I throw missiles that outpace the SAM
I leave in stealth
  and return in Black anger.
O---m! Ohhh-mmmm! O----hhhhhhmmmmmmmm!!!'

                           (1977)

During the early 1970s, the Black Consciousness Movement adopted an aggressive cultural policy in order to promote black liberation. Part of this policy was to use literature (especially poetry and drama) to increase social awareness and cultivate black pride.

Black Consciousness writers explicitly rejected the norms and values of Western writing. This is spelt out unequivocally by Mothobi Mutloatse in his introduction to *Forced Landing* (1980) – a collection of stories and essays. Mutloatse writes:

We will have to *donder* conventional literature: old-fash-ioned critic and reader alike. We are going to pee, spit and shit on literary convention before we are through; we are going to kick and push and drag literature into the form we prefer. We are going to experiment and probe and not give a damn what critics have to say... We are not going to be told how to re-live our feelings, pains and aspirations by anybody who speaks from the platform of his own rickety culture.

Artistic shaping, close attention to formal technique and the expression of meaning through indirect allusion were generally replaced by the unambiguous expression of bitter experience and accusatory anger. This was to be an aggressive literature that confronted political issues directly and forcefully. Works were no longer judged according to their form or stylistic complexity, but on the grounds of their political message. 'Good' literature was literature that promoted the struggle for liberation. Works that provided an authentic account of actual social conditions in South Africa or that succeeded in stirring their audience to adopt a more active political position were valued most highly. Increasingly, the Western canon of great literature was challenged by academics and by political activists. It became impossible to ignore the powerful voices of this new generation of writers and critics, or to assume a neutral position as the political situation in the country deteriorated. Even the ruling Nationalist government took notice and many writers were detained, imprisoned or exiled and their works banned.

We'll look at literature and politics in more detail in Chapter six. For now, though, we need to note that the rise of what came to be known as protest or resistance literature sparked a great deal of discussion. Some critics argued that politics had no place in literature, claiming that overtly political literature was sloganeering or propaganda. If literature made direct political statements, what was the difference between a literary work and a political speech? In addition, critics argued that there was no evidence to suggest that literature is capable of bringing about social change. Viewed from this perspective, protest writing not only diminished any proper sense of what constituted literary art, it was also impotent – a blunt instrument useless in confronting the might of the apartheid state. Many critics were disturbed by what they saw as the loss of artistic or aesthetic qualities. To their minds, protest poetry was undisciplined, unrestrained and erratic. Moreover, literature that blatantly stated its aim left very little room for conventional criticism. Finally, some critics were suspicious that one rigid system of value was being replaced by another system that was no less authoritarian. In the past, the criteria for evaluation may have been form or subtlety: now the criterion was the expression of a particular political viewpoint.

Supporters of protest writing offered a two-pronged defence. Firstly, they pointed out that all literature is in some sense political. If the term 'politics' is taken to mean the ways in which people organize themselves in society, then any work that deals with humans and their societies is political. Even works that deliberately ignore the conditions of social life often silently make a political statement. Accordingly, South African literary works that ignored the realities of South African life could be seen as making as much of a political statement as those that engaged directly with their context.

The second line of defence involved attacking the assumptions that held up British or American writing as an ideal against which local writing should be evaluated. Critics pointed to significant differences in context between South Africa and Britain or North America. They argued that the criteria of Western literary value were inappropriate to writing in a society marked by cultural diversity and deep political tensions. They believed that South African writing needed to be evaluated in its own right and on the basis of its own criteria. For these critics, insisting on foreign notions of what makes good writing was an act of colonization because the norms and values of the colonized society were considered inferior to those of the colonizer. Rejecting the criteria of the Western canon was an essential step towards establishing a new and relevant national culture.

It is important to note that attacks on the South African literature of the 1970s were not necessarily a sign of political conservatism or opposition to the liberation struggle. Almost as much criticism came from the radical left-wing as from the conservative right. Radical critics argued that even literature that actively promoted political liberation distracted from the actual struggle. Instead of encouraging people to take forceful action in opposing their circumstances, literature acted as a vent for anger. It became an outlet that dissolved tension and inhibited resistance. It was argued that the armed struggle, mass demonstrations, strikes and civil disobedience were the only appropriate responses to apartheid. Several critics also claimed that protest writing missed its proper audience. Written in English, it was inaccessible to the vast masses of the oppressed who, for the most part, were deliberately excluded from education by government policy and who spoke African languages. Protest writing, then, was consumed mainly by a liberal élite, and instead of inciting resistance it eased their consciences. Finally, particularly those critics who aligned themselves with Marxist literary theories questioned the political messages of protest literature. It was not enough to insist on human rights and democratic equality as was the case in most protest writing; literature should actively campaign for the restructuring of society according to socialist ideals of a working class revolution. All these arguments gained force as South Africa entered the 1980s and the literary establishment (writers and academics) seemed increasingly powerless against a regime that declared successive states of emergency and resorted increasingly to violent acts of suppression to retain its hold on power.

As we discussed in the previous chapter, history is never neat. Our account of the changing fate of South African literature suffers from the same problems as any historical narrative: it is a selective version of a complex series of events and ideas. This is as much a problem when discussing the present state of literary value in South Africa as it is when documenting the past. Nevertheless, this section will try to capture something of the literary climate of the 1990s.

Over the past decade, the 'art versus politics' debate outlined earlier has been replaced by the search for literary criteria that are more appropriate to the 'new' South Africa. Many commentators have argued that South Africans have to establish a sense of national unity while allowing for cultural and linguistic diversity. Similarly, the need for transformation in education, the economy and social life has influenced writers and critics as they search for new ways of expressing themselves. Finding a new voice is not an easy task, though, and there are varying opinions on the current status of local literary production.

Not surprisingly, the ruling African National Congress is in the most influential position because it has the power to determine the country's educational and cultural policy. In any society, the government influences art and literature by making laws and by subsidizing schools, universities and the performing arts. In addition, the South African Constitution, while guaranteeing freedom of speech, contains several clauses that have a direct impact on what may be said in public and, as a result, in literature.

One of the clearest statements of ANC policy was made by Albie Sachs at a seminar in Lusaka in 1990. In a paper titled 'Preparing Ourselves for Freedom', Sachs sketches a broad vision of an arts and culture policy based on what were at that time the proposed Constitutional Guidelines. Here is an excerpt from Sachs's paper.

In my view there are three aspects of the Guidelines that bear directly on the sphere of culture.

The first is the emphasis put on building national unity and encouraging the development of a common patriotism, while fully recognising the linguistic and cultural diversity of the country. Once the question of basic political rights is resolved in a democratic way, the cultural and linguistic rights of our diverse communities can be attended to on their merits. In other words, language, religion and so-called ways of life cease to be confused with race and sever their bondage to apartheid, becoming part of the positive cultural values of the society.

It is important to distinguish between unity and uniformity. We are strongly for national unity, for seeing our country as a whole, not just in its geographic extension but in its human extension. We want full equal rights for every South African, without reference to race, language, ethnic origin or creed. We believe in a single South Africa with a single set of governmental institutions, and we work towards a common loyalty and patriotism. Yet this is not to call for a homogenized South Africa made up of identikit citizens. South Africa is now said to be a bilingual country: we envisage it as a multi-lingual country. It will be multi-faith and multi-cultural as well. The objective is not to create a model culture into which everyone has to assimilate, but to acknowledge and take pride in the cultural variety of our people.

We do not plan to build a non-racial yuppie-dom which people may enter only by shedding and suppressing the cultural heritage of their specific community. We will have Zulu South Africans, and Afrikaner South Africans and Indian South Africans and Jewish South Africans and Venda South Africans and Cape Moslem South Africans (I do not refer to the question of terminology – basically people will determine this for themselves). Each cultural tributary contributes towards and increases the majesty of the river of South Africanness. While each one of us has a particularly intimate relationship with one or other cultural matrix, this does not mean that we are locked into a series of cultural 'own affairs' ghettoes. On the contrary, the grandchildren of white immigrants can join in the toyi toyi – even if slightly out of step – or recite the poems of Wally Serote, just as the grandchildren of Dinizulu can read with pride the writings of Olive Schreiner. The dance, the cuisine, the poetry, the dress, the songs and riddles and folk-tales, belong to each group, but also belong to all of us.

Each culture has its strengths, but there is no culture

that is worth more than any other. We cannot say that because there are more Xhosa speaking than Tsonga, their culture is better, or because those who hold power today are Afrikaans speakers, Afrikaans is better or worse than any other language.

Every culture has its positive and negative aspects. Sometimes the same cultural past is used in diametrically opposite ways, as we can see from the manner in which the traditions of Shaka and Ceteswayo are used on the one hand to inspire people to fight selflessly for an all-embracing liberation of our country, and on the other to cultivate a sanguinary tribal chauvinism. Sometimes cultural practices that were appropriate to certain forms of social organisation become a barrier to change when the society itself has become transformed – we can think of forms of family organisation, for example, that corresponded to the social and economic modes of pre-conquest societies that are out of keeping with the demands of contemporary life. African society, like all societies, develops and has the right to transform itself. What has been lacking since colonial domination began, is the right of the people themselves to determine how they wish to live.

The second aspect of the Guidelines with major implications for culture is the proposal for a Bill of Rights that guarantees freedom of expression and what is sometimes referred to as political pluralism. South Africa today is characterized by States of Emergency, banning orders, censorship and massive State-organized disinformation. Subject only to restrictions on racist propaganda and on ethnic exclusiveness such as are to be found in the laws of most countries in the world, the people in the South Africa envisaged by the Guidelines will be free to set up such organisations as they please, to vote for whom they please, and to say what they want.

We want to give leadership to the people, not exercise

control over them. This has significant implications for our culture work not just in the future, but now. We think we are the best (and we are), that is why we are in the ANC. We work hard to persuade the people of our country that we are the best (and we are succeeding). But this does not require us to force our views down the throats of others. On the contrary, we exercise true leadership by being non-hegemonic, by selflessly trying to create the widest unity of the oppressed and to encourage all forces for change, by showing the people that we are fighting not to impose a view upon them but to give them the right to choose the kind of society they want and the kind of government they want. We are not afraid of the ballot box, of open debate, of opposition.

All this has obvious implications for the way in which we conduct ourselves in the sphere of culture. We should lead by example, by the manifest correctness of our policies, and not rely on our prestige or numbers to push our positions through. We need to accept broad parameters rather than narrow ones: the criterion being pro- or anti-apartheid. In my opinion, we should be big enough to encompass the view that the anti-apartheid forces and individuals come in every shape and size, especially if they belong to the artistic community. This is not to give a special status to artists, but to recognize that they have certain special characteristics and traditions. Certainly, it ill behoves us to set ourselves up as the new censors of art and literature, or to impose our own internal states of emergency in areas where we are well organised. Rather, let us write better poems and make better films and compose better music, and let us get the voluntary adherence of the people to our banner ('it is not enough that our cause be pure and just; justice and purity must exist inside ourselves' – war poem from Mozambique).

Finally, the Guidelines couple the guarantees of individual rights with the necessity to embark upon programmes of affirmative action.

This too has clear implications for the sphere of culture. The South Africa in which individuals and groups can operate freely, will be a South Africa in the process of transformation. A constitutional duty will be imposed upon the state, local authorities and public and private institutions to take active steps to remove the massive inequalities created by centuries of colonial and racist domination. This gives concrete meaning to the statement that the doors of learning and culture shall be opened. We can envisage massive programmes of adult education and literacy, and extensive use of the media to facilitate access by all to the cultural riches of our country and of the world. The challenge to our cultural workers is obvious.

✪ Do you agree or disagree with Sachs's arguments? Why?
✪ What criteria of literary value do you think are most appropriate for South Africa today?
✪ What similarities (if any) are there between Sachs's criteria for literary value and those expressed by the reviewers of *Disgrace*?
✪ To what extent do Sachs's criteria link with the ANC's evaluation of Coetzee's novel as racist?
✪ If you had to draw up a list of literary works that you think are essential reading for South African students, what works would you choose? Why?

· ·

SEVEN

WRITING

REVIEWS

REVIEWING IS PART of the literary institution and the most public form of evaluation of literary works (as well as plays, films, art and photography exhibitions, CDs and videos). Reviews serve a dual purpose: on the one hand, they give us some idea of the content of the work and, on the other, they judge the merit of the work, either recommending it or not. Thus, like all literary criticism, reviewing involves both interpretation and evaluation.

Reviewers and critics often use different criteria and often arrive at contradictory assessments: one reviewer may warmly applaud a work, while another may dismiss it. Sometimes one reviewer uses different criteria in the same review, something that can result in contradictory assessments of the work. In order to understand the basis on which a reviewer or critic lauds or rejects a work, we need to establish what criteria have been used and whether they are appropriate to the work.

 An example of a review

A brave journey on to new ground is weighed down by old baggage

Review: Tony Morphet

THESE THINGS HAPPEN
By Shaun de Waal
(A.D. Donker)
R66

TONY MORPHET writes widely on contemporary South African cultural studies, focusing in particular on the formation of cultural discourses.

1   *These Things Happen* is Shaun de Waal's first collection of stories. Some of the work has already appeared in anthologies but most of the 12 stories are new. The thematic centre of gravity in the collection is the conditions and experience of gay life, but this is counterbalanced and extended by an unexpected 'political' interest and a wider discussion about the nature of art.

2   The best story, 'Isn't He Dead', is a poised and subtle account of an old jazz pianist, a veteran of the early 60s, left behind and forgotten in London in miserable

circumstances, now suddenly called out by the 'movement boys' to play the atmosphere for a Swedish donor. The old man, Jake, is able to mix a wonderfully wry view of the newly rich and powerful politicals with an intense inward sense of the meaning of his art – for himself and his country. It's a nice irony that after the event the Swede wants to put money into recording the music – not into the movement.

3   'Isn't He Dead' shows something of what De Waal can do, as, in different form, do 'Dave' and 'Double Strands', two of the gay stories in the collection. In 'Double Strands', Oliver, a South African poet, is on a trip back home to be with his family as his father dies. He writes daily (in diary fashion) to Andries, his lover in London. The tension that shapes the story is between desire and love. Dad is repellent – the family likewise – while Andries represents everything that cultivated style can offer. Nonetheless, Oliver's new book will be dedicated to Dad.

4   In 'Dave', desire drives the narrator to try and find the actual man posed in his favourite gay pic. They get to meet but it turns into a cruel and sad defeat. It's only the money that the real Dave wants, and all that is left, when he is finally got rid of, is ash on the carpet and the original tantalising photograph.

5   De Waal's central interest is in the contradiction between the intense and endless interior flow of desire and the inevitable defeat it has to suffer in its attempts to reach fulfilment. It is not an easy theme to control. Desire subverts every structure; and structure thwarts the formlessness of yearning. Not surprisingly, questions of design emerge as a literary worry.

6   In 'X and I', written in memory of Koos Prinsloo, he has Prinsloo recall that, in writing, 'the structure, the design, always came to him first' – while the writer-figure in the

same story, sitting at his PC, wonders of his work, 'Is it too formless? Will it seem too personal?'

7 Politics offers one means of approaching the problem. 'The Stalwart' is about an ageing white SACP returnee who is taken on a tour of his old haunts by a young journalist. The narrative turns when the old man picks up that the journalist is gay and slowly reveals his own sense of loss and regret that he 'gave up' on loving a man in order to devote himself to the Party. The choice for structure cost too much. He closes his tour by fusing the old struggle with the new one for gay rights. The baton is passed on – but not even gay liberation will resolve the issue of desire. It too is 'structure'.

8 'Red Rover' also sets a political framework around the central issue. Roberts (an obvious stand-in for Ronnie Kasrils), recalling the lead-up to the charge at Bisho, poses the urgency of the crowd's desire for freedom against the blade-wire facing them. Catastrophe follows.

9 The political constructions are set up but do not resolve the literary problem. In several of the stories the author finds himself on both sides of the desire/closure contradiction – yet unable to allow it to play itself out. He relishes the open-ended urbanity of the gay milieu but he wants a holding frame as well. Too often the poise disintegrates and the stories drift.

10 For this the reader has to pay a price. Stranded between unfulfilled alternatives it becomes hard to gain a firm purchase on what might be at stake in the drama of consciousness and the events of the stories. Repeatedly, in the absence of a strong internal design, one finds oneself left subject in the middle ground to what feels like a manipulative purpose in the writer.

11 'Isn't He Dead' provides the comparison. It is sure and steady in the angle that it sets on the events and the forms of awareness it describes and it delivers a firm and

full judgement on the relation between the different fulfilments of art and politics. Design and texture work together – not in opposition.

12 De Waal is breaking new ground and the risks are substantial. The collection shows that he is especially vulnerable on two sides. One is the appeal of an unconstrained "telling it as it is"; the other is the security of fitting experience to a political pattern. Neither of these dangers is new to South African fiction writing and we have more than enough evidence of their baleful consequences. His ground is new but his procedures carry too much of the old baggage. Because they often lack a depth of resolution we miss the radically new insight into the people and the milieux that interest him.

13 Fortunately, as the best of the stories show, he is aware of the problems and he has the capacity to work his way beyond them. Nonetheless, he will face some difficult choices in the future.

TIME TO WRITE

Answer the following questions concerning Tony Morphet's review:

✪ What function does the first paragraph fulfil in the review?
✪ What do paragraphs two, three and four have in common?
✪ In what way is paragraph five different from those that precede it?
✪ How do Morphet's comments in paragraph five inform his discussion in paragraphs six, seven and eight?
✪ What function do paragraphs nine and ten perform?
✪ Why does the reviewer discuss one more story after paragraph ten?
✪ What conclusions does Morphet come to in paragraphs twelve and thirteen?

✪ Pick out some of the interpretive statements in the review.

✪ Pick out some of the central evaluative statements.

✪ What are the criteria of evaluation used in this review? Are they appropriate?

✪ Can you list some of the reservations that Morphet has about the collection of stories?

✪ What, in the final analysis, is Morphet's evaluation of *These things happen?*

YOUR OWN REVIEW!

By now you have encountered an array of positions on both literary interpretation and evaluation. You have also considered, in some detail, a selection of reviews of texts. We would now like you to write a review of a book that you have recently read. If you like, you may read *Disgrace* by J.M. Coetzee, or another South African text and engage in the critical debates on which this chapter has been based. You are, though, free to choose any book about which you have something to say.

Use this checklist as you write your review:

Have you:	Yes/No
• given the details of the text (title, author, year of publication, place of publication and publisher)?	
• suggested a possible interpretation of the text (what the significance is of important aspects of the work)?	
• explained something of the content (probably in the form of a brief plot synopsis)?	
• justified your reading of the text by giving evidence?	
• commented on the importance of the text?	
• decided on the merits of the work?	
• explained the basis on which you have evaluated the work?	
• quoted from the text to give a sense of the style of the work?	

PART TWO

Love, sex, body

OVERVIEW

In this part, we are going to explore representations of sexuality, the body and desire. These are both fundamental and ever-present in our society. Over the last fifty years, there has been an explosion in the extent to which these aspects of human existence have come to dominate popular culture. The media constantly bombard us with images of the body, sexuality and desire. Movies, billboards, television programmes, advertising and pop songs (to name only a few kinds of 'text') tap into the general fascination with these topics. It is impossible to escape being affected by this net of images and meanings. Representations of love, sex and the body have become so pervasive that we tend to accept them unquestioningly as part of our daily experience.

This fascination with sexuality and the intimate details of life is not new. Michel Foucault, whom we met in the previous part, traces modern Western attitudes to sex back to the nineteenth century, a time that we think of, incorrectly, as being conservative and repressed. History also tells us that many cultures at different times have had particularly liberal attitudes to sexual behaviour. Even within a single context, we can trace changing views on the nature of love, the morality or immorality of sexual behaviours and what constitutes a desirable body. A selection of literary texts or artworks from different periods will clearly show that ideas about love, sex and the body are dynamic.

Contemporary Western society is unique, though, not just because of the apparent openness of its attitudes, but also because of the immense power of the media to promote specific ideas about sexuality, the body and desire. Most of us are exposed to thousands of messages about how we should look, love and behave each day. Some of these messages are transparent in their exploitation of basic human desires, but many work subtly. We may not always be conscious that our ideas and attitudes are being targeted in deliberate and carefully calculated (often profit-making) campaigns. To a significant extent, our views on love, sex and the body are subject to careful manipulation by the media, rather than being a matter of personal choice, as many of us might like to think.

Interestingly, our vulnerability to the media comes at a time when people are increasingly aware of the sexual abuse of children, and of violence, especially against women. The media's fascination with sex is also being played out at a time when many societies, including South Africa, are reeling under the devastating impact of HIV/AIDS. Contrary to ideals of romance, we know that sex has the potential to destroy lives. In this context, the persuasive use of ideas about love, sex and the body, whether in the media or in literature, deserves close attention.

Our interest in this topic is critical and analytical. We are not advocating particular lifestyles or challenging specific moral, ethical and religious beliefs but rather exploring images of the body, love and sex in order to find out how they participate in prevailing and shifting ideas about men, women and their interactions.

In the first chapter of this part, we examine advertising and, in the second, we look at love poetry. The persuasive power of discourse is crucial to both chapters. Advertising is, on the surface, an attempt to persuade the consumer to buy a product. Love poetry, on the other hand, while celebrating love and lovers, also tries to persuade its addressees to respond in specific ways. By using what we have learnt about interpreting and evaluating in *Critical conditions*, we look at how the popular media and literature deliberately draw on ideas about love, sex and the body in order to achieve particular effects. We are also interested in how our personal views about these matters may be shaped by the books we read or the media to which we are exposed. As we uncover the ideas behind these different kinds of text and the codes they draw on, we can empower ourselves to be critical consumers rather than passive recipients of the messages that target us.

3 Bodies that sell

IN THIS CHAPTER we will be applying the skills of interpretation and evaluation to texts other than literary ones. We start by looking at the similarity between literary texts (mainly poems) and messages that we confront on a daily basis when we watch television, listen to the radio, or read magazines and newspapers. We then move from a general discussion of the media to a particular instance: magazine advertising. Since magazine adverts are themselves diverse texts, we will be looking at one aspect in particular, namely the use of bodies, desire and sex in advertising.

Literary texts are forms of communication that seem to demand interpretation. At school and university we are taught to analyse literature and to learn the strategies of literary criticism. We expect that reading literature critically will reveal 'deeper meanings'. We also tend to believe that an awareness of contextual and stylistic matters will sharpen our insights.

Our responses to the media seem to be more passive and less detailed. The messages that we receive from movies and magazines, television and pop music are arguably more powerful and have a greater capacity to influence a mass audience than those in literature. Yet we seldom subject these messages to the same scrutiny as literature – even though we should. The techniques of literary criticism can be extremely effective when applied to the media. Like literature, the media fit into the communications model that we used for discussion in Chapter one:

SENDER ------------- MESSAGE----------RECEIVER

In the case of advertising, the sender is the company or person advertising a product or service. The message is the advertisement and the reader (usually a consumer) is the target audience. In the same way that literature relies on the sender and reader sharing language, so advertisers manipulate visual and linguistic codes that they know the public will understand and respond to.

ONE

THE MEDIA AND MEANING

THE MODERN WORLD is dominated in many important ways by the mass media. The 'mass media' are the means by which messages are sent to thousands, even millions, of people simultaneously. Just as a literary text is a medium, the popular press (including magazines), television, radio, film and the Internet are media. They are the means by which messages, such as radio and television programmes, magazine and news articles, movies, and all forms of adverts, are communicated to people *en masse*. Whether consciously or unconsciously, we are continually involved in interpreting messages relayed to us by the mass media.

THE GREAT MEDIA DEBATE

The debate about the extent to which we, as individuals and as a society, are affected by the mass media is a lively and interesting one. Many people think that the frequent representations of violence in films and on television lead to increased violence in society. Similarly, many people criticize advertising on the grounds that it promotes certain lifestyles, body-shapes and attitudes over others. This can be detrimental when people imitate what might be inappropriate. Most agree that the influence of the media on people's day-to-day reality is not simple. Watching violence does not lead directly to people enacting violence.

Being exposed to the advertising ideals of thin bodies does not lead directly to anorexia. We cannot, though, deny that the media affects our experience of reality. Any society has a set of ideas that is dynamic and powerful, and that influences both how that society functions and how people think of their own identities. We term this the 'ideology' of a society. Clearly the mass media are primary sites for the production of ideology: they feed into our understanding of ourselves and our world in complex but important ways.

- How do the media affect your priorities and beliefs?
- What negative effects do the media have on society?
- Can the media be used for the good of individuals and communities? If so, how?

LOVE, SEX, BODY

⬤ The media and you

Fill in the following table and add up the number of hours you spend engaging with different forms of mass media.

Medium	Hours
How many hours a week do you spend watching television?	
How many hours a week do you listen to the radio?	
How many hours a week do you spend reading newspapers or magazines?	
How many hours a week do you spend watching movies?	
How many hours a week do you spend on the Internet?	
Weekly Total	

It is interesting to reflect on the extent of our involvement in mass communication. Many such engagements occur without our being aware of interpreting and evaluating messages. However, every message we encounter involves making decisions about its meaning, and we regularly decide whether a film, magazine article, advert, or television programme is good or bad according to our own criteria. We might struggle to explain those criteria if we were asked; often we feel simply that we enjoy something and therefore experience it as 'good'.

Because of the amount of time that we spend at the receiving end of the mass media, and because of the power such communications have over us, many people are interested in analysing the ways that the media construct messages. This is reflected in the development of key aspects within the growing academic disciplines of Communications (including Media Studies) and Cultural Theory. These fields of inquiry are concerned not just with the construction of individual messages (how they are put together), but also with the ways in which the media both reflect and create our society.

We have seen that we may be very aware of the act of interpreting literature because of the difficulty of the task. Because we interpret literary texts consciously, we are aware of being engaged in the activity of making decisions about meaning. Our central question in this chapter is: do we interpret messages such as adverts, articles in newspapers, music videos, episodes of our favourite television programmes or movies in a way that is similar to our interpretation of poems?

Think back to the detailed analysis of Sue Clark's 'Biting Through' in Chapter one. Now imagine reading a magazine article about an interesting topic. Obviously we do not sit, pencil in hand, breaking down the magazine article into individual images and finding out how they relate to one another. However, we do give a meaning to the article. It communicates a message to us and we are therefore involved in an act of interpretation. Furthermore, we might tell a friend that the article is 'interesting', 'well written', or 'badly argued'. These comments suggest that we have evaluated the article. It is possible to argue that we more readily evaluate the popular media than we do literature: we are more prone to say, 'That was a good movie' or 'This is a good magazine' than we are to claim, '*Heart of Darkness* is a good novel'. As soon as we approach a text as a 'literary' construct, we tend to think through matters of meaning and value far more carefully than when we come across other messages. If this is the difference between literary and other messages, we need to ask what the difference is between writing a poem and designing an advert.

Poems and adverts are both messages that are carefully constructed to have certain effects on the reader. A wide range of decisions is made in the course of constructing the messages. Just as the poet combines words and phrases that have particular

meanings, so the person or group of people designing a magazine advert choose images and words that combine to mean
something specific or to have a particular effect on the reader. A
magazine advert, just like a poem, is a message communicated by
means of a medium and it, too, is carefully crafted and can be
analysed to see both how it is constructed and the effect it is
designed to have.

From this brief discussion we can see that messages relayed by
the mass media do warrant analysis. We cannot simply take them
at face value and, although we are not often aware of it, we are
involved in a constant process of interpretation and evaluation as
receivers of those messages.

- Is there any similarity between writing a poem and designing an advert for a magazine?

What is a 'text'?

What is a 'text'? Here is *The Oxford Advanced Learner's
Dictionary* entry:

> **text** /tekst/ *n* **1** [U] the main written or printed part
> of a book or page, contrasted with notes, illustra-
> tions, etc: *too much text and not enough pictures* ○
> *The index refers the reader to pages in the text.* **2(a)**
> [C] the written form of a speech, a play, an article,
> etc: *the full text of the President's speech* ○ *As an
> actor, I keep going back to the text when I'm working
> on my performance.* (**b**) [U] any form of written
> material: *a computer that can process text.* **3** [C] (**a**)
> a book, play, etc studied for an examination: *'Ham-
> let' is a set text this year.* (**b**) a piece of writing
> about which questions are asked in an examination
> or a lesson; a passage(6): *Read the text carefully and
> then answer the questions.* **4** [C] a sentence or short
> passage from the Bible, etc used as the subject of a
> SERMON(1) or discussion: *I take as my text...*

You will notice in the definition that the emphasis is on the written part of a communication, as opposed to pictures or a text
spoken aloud. This definition is all very well, but the word 'text'
has come to mean something a little broader in both the study of
literature and the study of the mass media. We now use the term

to refer to the combination of words and images that make up a given message. Phrases such as the 'advertising text', for example, refer to the combination of all the elements that contribute to the meaning of the advert.

According to literary theorists, the reason for this change (from 'text' as written to 'text' including all elements that communicate meaning) is that, just as words get their meaning from the languages of which they are an element, all pictures, gestures and objects derive their meaning from their own systems of communication, from their own language. There is, for example, a language of pictures in which each image (or part of an image) has a certain place. Every feature of a text has meaning because of its relationships with other elements in a system of communication. Therefore, meaning exists in complex systems known as codes. In order to construct a message, we use codes and assume that the code is shared (at least to some degree) by the person who will 'read' the message. Codes of meaning extend beyond language into the visual (and into all other ways in which meaning is communicated).

These ideas form the basis of theories of meaning that we group under the broad field of semiotics. This area of study is derived from structuralism, which we mentioned in Chapter one. According to semioticians (such as Roland Barthes, the French philosopher, and Umberto Eco, whom we will be discussing later), the meaning of any single element (a word, an image, a symbol) can be understood only once we have some idea of all the elements of its set. We need to understand the code that gives the individual element its meaning if we are to make sense of it.

CULTURAL CODES: THE CASE OF THE CRUCIFIX

Imagine being confronted with a picture of a crucifix (a cross bearing the crucified body of Jesus Christ). Most people, even if they are not Christian, know something of the system of meaning that gives the symbol significance (both its meaning and importance). They might understand this significance by likening the symbol to something holy in their own culture. What would happen, though, if a crucifix were handed to a group of people who knew nothing of Christianity? They would know nothing of the system of meaning of which the symbol is a part. For them, the meaning might not relate to its religious significance at all. They might think the object purely decorative or might relate it to something completely different in their own culture.

The crucifix itself does not have meaning. Its meaning derives from the Christian account of Christ's death and its importance. Without knowing the system of meaning, and without the object being different from others in that system, it is simply meaningless. We need to understand the code of Christian meaning to interpret the crucifix meaningfully.

Let's look at an example.

✪ Would each of these women be appropriately dressed in cultures other than their own?
✪ Do these clothes have particular meanings in the cultures from which they derive?
✪ Do these meanings change when the clothes are worn in other cultures?
✪ Would it be acceptable for a woman to adopt the traditional dress of another culture because she likes the 'style'? Explain your answer.

Meaning, it seems, is found within codes of significance and it varies between (and sometimes within) cultures. In order to interpret, we need to consider the code that generated the meaning, and reflect on the code we are using to decipher the 'text'.

But is it 'real'?

This idea of meaning (namely, that things acquire meaning

through codes) raises the question of what is 'real'. Surely many things have simple meanings that are shared by all codes. There must, we usually assume, be a 'real' world where meanings are fixed and cannot be disputed. How is the real world reflected in and by the media? Do the media simply mirror reality?

TIME TO DISCUSS

To explore the relationship between the 'real' world and the world of the media, try answering the following questions:

- To what extent does television influence clothing fashions?
- How does what we watch on television or read in magazines affect the way we think of ourselves (our nation, our race and our bodies, for example)?
- In what ways might television change the way people speak and behave?
- Which countries, or cultures, dominate the mass media and what effects does this have?
- To what extent, given your answers so far, do the media make the world in which we live?

The relationship between the world represented in the media and the 'real' world is rather more complex than we might expect. It is almost impossible to exist in the world without feeling the influence of the mass media. Here are three arguments relating to this relationship. Decide in each case whether or not you agree and what reasons you would give for your agreement or disagreement.

ARGUMENT 1

Almost all television programmes rely on advertising for their economic survival. Advertisers will favour those stations (or television networks) that are watched by most people because this will ensure a larger audience for their adverts. Therefore, television stations will tend to make decisions on the basis of what is popular and economically profitable rather than what is true, accurate or valuable.

LOVE, SEX, BODY

Adverts on television or in glossy magazines (such as *Cosmopolitan*, *Ebony* or *Bona*) try to sell us a particular product. They do so by suggesting that we can be different people (look different, live different lives) if we buy the product. But as we know, the things we buy cannot really alter our lives. Therefore, all advertising is based on dishonesty.

Although the world is represented on radio and television as if it is a reflection of the 'real' world, no news item is a neutral mirror of the world: any news item involves the selection of a story (from millions of possible stories) and the choice of a particular way of telling the story. However, we are likely to confuse the 'real' with what we see on television or listen to on radio because we forget about the selection process and are often influenced by the way the story is told.

Each of these arguments draws our attention to important aspects of the media. Any media message is constructed through a process of selecting facts and ways of presenting them. This selection is motivated by various concerns, such as popularity and the economic and political interests of broadcasters and their sponsors. Precisely because the news, for example, is *made* and not simply shown, we cannot assume a direct equivalence between the world 'as it is' and the world as it is reflected in news broadcasts. We cannot simply accept that what we see and hear in the media is the only way that things could have been represented. It is useful to study the media to examine how messages are constructed, what effect they are meant to have on us, and why that effect might be sought by particular people or groups.

A DIFFICULT QUESTION: REALITY VS. THE 'REAL'

The word 'real' has sometimes been placed in inverted commas in this section to show that reality itself cannot be reached except through representations of it. That which we accept as 'real' is simply the most common representation of the world and ourselves. What seems natural to us ('that's just the way things are') is simply a set of representations that we have forgotten are representations. It is as if, in some fundamental way, we are inside a story that we have forgotten is a story. The media is a primary site for storytelling.

If we accept that we regularly confuse reality with the world represented in the media, can we know what is actually 'real'? Surrounded by so many representations, how can we hope to tell the 'real' (as it is represented) from reality?

IN SUMMARY

LET'S CONSOLIDATE OUR understanding of the media:

✪ All media messages (radio and television programmes, articles in newspapers and magazines, all adverts, even billboards) are communications with an audience.
✪ The media select and represent specific versions of reality.
✪ These communications are carefully constructed to achieve certain effects.
✪ In the course of construction, codes of meaning are used (either language or the 'language' of visual images).
✪ We cannot assume that media messages are literally true or that they are accurate reflections of the real world (the world as it is).
✪ We can analyse media texts to see how they are constructed and to see what interpretations of those texts are possible and informative.

LOVE, SEX, BODY

The media select and represent specific versions of reality.

. . . .

WE ARE SURROUNDED by advertising. It is, quite deliberately, inescapable in the modern world. We encounter advertising on radio and television, in magazines and the press, plastered on billboards in our cities and on every Internet homepage. It has been estimated, for example, that Americans encounter between 300 and 2 500 advertisements of different sorts each day. Advertising is a form of communication motivated by a specific intention. Adverts are meant to get consumers to do something – to buy particular products or use particular services. Some are intended to get us to vote for particular political parties, or to behave more responsibly in society. We can learn a lot about

TWO

'READING' ADVERTS

communication and about the world in which we live by analysing the ways that adverts seek to achieve their aims.

Adverts are texts. They combine images and words (and sometimes sound) to communicate a message about a particular product or service. Some adverts simply give us information, but most use a range of manipulative or persuasive techniques. One of these is to link the product or service being sold to something other than itself. This is a key issue and we will explore it in detail as we look at advertising.

 Creating your own advert

Congratulations!

You have just been appointed as an advertising executive and designer.

THE DIRECTORS OF a small company have asked you to help them market a new product. You have to write down what you, in your capacity as an advertiser, plan as a campaign for their product. Don't panic. You will be guided, step by step, through the process. The left-hand column contains tasks for you to do. The middle column gives you examples that might be used in a similar campaign, though for a different product. The steps for the parallel campaign are shown with a hand icon (✍). Wherever a pencil icon (✐) appears, you have to make a decision and write down what you plan to do.

YOUR CAMPAIGN	PARALLEL CAMPAIGN	NOTES
THE PRODUCT ✐ You have to design a campaign to sell a new carbonated (fizzy) soft drink.	✍ I will be designing a parallel campaign for a new chocolate bar with nuts.	

SPECIFICATIONS

✐ The drink is a yellow fruity drink with a lemon-based flavour. You are free to add vitamin additives if you would like to. You can choose to package it in cans, bottles or both.	✐ The chocolate bar is made up of five square blocks of milk chocolate and includes hazel nuts. It is a suitable snack for one person, and is packaged in foil.	

Step 1 The Target Market

✐ No product can be marketed to everyone equally. You will have to decide on a specific target market. Which section of the community do you wish to focus on and attract in your campaign?	✐ I want to market my chocolate bar to young people (say between the ages of eleven and nineteen) who are conscious of health and fitness; those who see themselves as 'sporty'.	

Step 2 The Theme

✐ This is the difficult moment. You now have to think of a *concept* that will link the aspects of your campaign. What overall image do you want your product to have? To what central idea will you relate the various components of your campaign? How does this theme or concept link to your target market? You might like to do some marketing research to try and uncover the ambitions and desires of your target market.	✐ Advertising campaigns entail choosing a theme or *concept*. All aspects of a campaign relate to this concept in one way or another. You need to create a product image that is regularly reinforced and that will stick in the minds of your target market. Since my market is young, health-conscious adults, my concept might be sporting excellence, linking the idea of the chocolate to energy and athletic performance.	

YOUR CAMPAIGN	PARALLEL CAMPAIGN	NOTES
Step 3 The Name ✐ Taking account of your concept and target market, you now have to name your drink. Think of something that is memorable and catchy. Alliteration is often useful for slogans and names.	✐ I want a name that is 'catchy', that will stick in the mind of my targeted consumers. The name should relate to *sporting excellence* in some way. My chocolate bar is to be called *Expressive*.	
Step 4 The Slogan ✐ It is now up to you to formulate a slogan for your drink. A slogan is a statement of no longer than about eight words, which you will use in all parts of your campaign.	✐ A snappy slogan is the basis of most successful advertising. Slogans are commonly repeated in print and in television and radio adverts. My slogan is: *Expressive – you'll finish first in more ways than one.*	
Step 5 The Image and the Words ✐ Now design a magazine advert that incorporates your slogan and the product name. You can actually draw the advert (or use parts of images from other adverts). The advert must use both a message and pictures that will appeal to them. In the text, you might like to employ some poetic techniques. Puns, alliteration and rhyme work effectively to make the advert more memorable.	✐ In designing a magazine advert, the advertiser needs to combine images with words in a way that entices the potential potential buyer (the target market). I will use a photograph of a hurdler scaling a hurdle. Her feet are blurred, creating an impression of speed. Below the image is a photograph, in high colour, of two *Expressive* bars and the slogan appears, in white print, at an angle (from bottom left to top right) across the image.	

YOUR CAMPAIGN	PARALLEL CAMPAIGN	NOTES
Step 6 Radio and Television	✍ In keeping with the magazine advert, I would make use of hip-hop music and a number of fast-changing montage shots of great sporting moments: hurdlers, long-distance runners, basketball players and soccer players. The advert would then cut to the locker rooms of the athletes, where they would all be eating *Expressive* bars.	
✐ Without worrying too much about the technicalities, sketch out (roughly) an idea for a radio advert and a television advert. What would you include in these? You should also consider imaginative ways of using your slogan and the name of the product. Remember that you are now able to use the medium of sound.		

TIME TO REFLECT

BEFORE YOU REFLECT on the campaign you designed, let's think about the one that I proposed. I have suggested the use of an athlete (the hurdler) to advertise a chocolate bar. There is no logical connection between my concept and the product being advertised. Obviously chocolate bars do give you energy, but they are not directly associated with sport. Eating an *Expressive* bar will not make one a better athlete, and certainly not a fit and capable hurdler. My campaign is designed to create an association, in the mind of my target audience, between the chocolate bar and fit, healthy bodies, which are necessary for sporting success. The link the campaign creates is a constructed and not a natural one. It is not true or literal on any level. As an advertiser I hope that my prospective consumers' desire to be sporty, healthy and successful will entice them to buy the chocolate.

Think about your advertising campaign and answer the following questions:

○ What activities or lifestyle did you associate with your drink?
○ Is there any logical connection between the drink and your associations?

✪ Are you appealing to a particular desire, need or fear that consumers might have? If so, what is it?

Adverts: intentions and methods

In Chapter one we discussed intentionalism. The intentions of the author are relevant in interpreting literary texts, but are difficult to establish because they might be complex and multiple, or they may not be fully conscious. An author's intentions may or may not be achieved in the course of writing. Sue Clark's comments on her intentions in writing 'Biting Through', particularly the connection between her poem and its political context, are a case in point. When we come to thinking about the intention of adverts, though, the matter is simpler. Adverts are formulated with the specific and conscious intention of selling something or of persuading people to do something, such as wear condoms or vote for a particular political party. They are also intended to have you use a product or a service at a rate and in quantities that benefit the manufacturers or providers. Not only do they want you to buy a product once, they also try to make the product seem an indispensable part of your life so that you will buy it repeatedly. We need to examine the methods advertisers use to achieve these aims.

Adverts often connect things that are not logically connected. They do so in order to establish a relationship between a product and some desire (something the consumer wants or wants to be). My potential target market is meant to link youthful fitness, fashion, excellence and sexiness with the *Expressive* chocolate bar. Creating a taste for a product and tapping into the desires of consumers is at the heart of advertising techniques.

Imagine a world in which adverts simply reflected reality. Let's take cigarette adverts for example. (Before you begin this exercise, you may need to remind yourself of what cigarette adverts look like by looking at billboards and in magazines.)

✪ Imagine that the world is as it is represented in cigarette adverts. What kind of world would this be?

✪ What activities would people participate in?
✪ How would people dress and would they all be beautiful?
✪ What age would people in our cigarette-world be?

Clearly advertisers try to link smoking with particular lifestyles, bodies and activities. While perhaps no one believes that smoking will literally (or actually) make them young, rich and beautiful, a successful outdoor adventurer or a competent cowboy, a desire to be a certain type of person is linked to a particular brand of cigarettes. In this way a desire to be different is connected with a particular product. Also, the consumer does not have to 'believe' adverts in order for them to be effective. Just as our desires are complex and often hidden (even from ourselves), so advertising relies on a link between desires and products that is not fully conscious natural, or logical.

There is an obvious difference between the lives of ordinary people and the lives portrayed in adverts.

We can summarize these ideas by saying that there is a gap or discrepancy between the real world and the world of desires and imagination. Advertising functions in this space. Stated simply, for most people there is a significant difference between what we have (or are) and what we want (or want to be). While some see this difference as a fact of existence, others might claim that the difference is created, at least in part, by the media messages that surround us.

TIME TO WRITE

BEFORE WE MOVE on to a detailed discussion of bodies and sexuality in adverts, let's analyse a magazine advert in terms of our discussion so far. Study the Clique advert below and answer the questions that follow.

NOTES OF BEAUTY by Clique
The **Clique** range of cosmetics is available exclusively at selected **Clicks stores** nationwide.

Photographer: Kevin Fitzgerald
Model: Lisa Cowley from The Model Company
As seen in *SL* magazine, September 1998
Reproduced with permission from *SL* magazine.

✪ What range of products is being advertised?
✪ Does this advert appeal to you?

✪ Who do you think would be the target market for this advert?
✪ What is the connection between the slogan ('Notes of Beauty') and the images?
✪ What associations does classical music have in society?
✪ In each of the pictures, to what specific part of the image is your attention drawn?
✪ What forms of desire, as discussed earlier, are being used in this advert?

Let's begin by considering the selections made by the designer of the advert. The model is not (in all likelihood) a musician. She has been selected for reasons other than her musical talent. Her complexion, eyes, lips and hands represent conventional Western ideas of beauty. At the same time, they show the cosmetics to their best effect and create the sense that the range of cosmetics has in some way created her appearance. We are led to believe – consciously or unconsciously – that she is beautiful because she is wearing the particular lipstick, eye shadow, nail polish and mascara that are being advertised. Given that the model represents a socially accepted ideal of beauty, we can assume that the advertisers intend to link the desire for this ideal to their product. The implied message is: 'You, too, can look this good (and enjoy all that follows from being beautiful) if you use this range of products.' We also have no guarantee that the cosmetics used to create the impression of the model's perfection are the ones being advertised. She has been made up by beauticians who will obviously use every means and product at their disposal to achieve the desired effect. The consumer, reading the advert, automatically assumes that the advertised products have been used, but we have no way of knowing this.

The next question is: Why the musical instrument? In our society, classical music has certain associations. In Chapter one, we discussed the idea that words have connotations. We link certain ideas with words; words evoke associations in our minds. Objects and activities also have connotations. The viola implies sophistication. It is associated with classical music, an activity generally considered 'cultured' and enjoyed by a privileged sector of society. (The connotations of an electric guitar, for instance, would be entirely different and inappropriate, given the concept and target market of the advert.) The instrument, then, connects the cosmetic range with a particular set of meanings. Here is a list of words that spring to mind when I think of the way in which the instrument is used in this advert (all of which can be looked up in your *Oxford Advanced Learner's Dictionary* if you are unsure of their meanings): romantic, lyrical, harmonious, refined, tranquil, thoughtful, seductive, sophisticated, contemplative and alluring.

From this brief analysis of just two aspects of the advert (the model's body and the musical instrument), we can see that advertising relies on certain existing connotations (like those of classical music) and makes use of certain socially accepted (although not 'natural') ideals, in this instance, of feminine beauty. Thus, the advert constructs a message from elements that already have meanings because of the way they are encoded in a particular culture. Advertising, like the construction of a literary text, involves selecting elements with a range of meanings and combining them to achieve specific effects. Adverts thus use existing meanings, but combine them to create distinct messages.

One could claim that for a long time cosmetics have been associated with norms of femininity and that wearing make-up (of some sort) has become a standard practice in most societies. However, it would be wrong to claim that people have a natural desire for a particular range of products. Advertising takes general desires (to feel good, to attract the affirmation of others, to be physically gratified) and attaches them to specific products. It manipulates our wants and needs by connecting them to specific items or activities. To appeal to someone's desires in an advert is to use something that already exists and to link it to something you wish to market. In doing so, adverts both use and redirect desires.

Thus far we have overlooked perhaps the most obvious aspect of the Clique advert: a woman's body is being used to sell something. The model is turned into an object of desire. She is sexy, in an understated way. For the rest of this chapter, we will focus on the use and 'creation' of bodies in advertising. We will explore the idea that 'sex sells' by examining the ways in which sexual desire is manipulated to market particular products and services. Not only does advertising offer an interesting range of texts that allow us to understand some of the ways in which bodies are made and used in our society, but its strategies and methods are also used in other popular media. Understanding a code of meanings from one context helps us to think about others more clearly.

. . . .

WE HAVE CONSIDERED human desires in quite general terms up until now. The rest of this section concerns a particular range of desires, those linked to our sexuality. Advertising often links products and services to our sexual desires, insecurities or fears. In this process, our sexuality is not only reflected, it is also created in complex, but centrally important ways.

THREE

BODIES, SEX AND CONSUMERS

Is sex really so important?

Why all this fuss about bodies and sex? People are often surprised by the emphasis placed on sexual desire in analyses of literary (and cultural) texts. What often concerns people is that these analyses seem to reduce the complexity of human experience to one aspect. Sexuality is closely linked to our identities: the desires we experience, and our sexual needs and fears, are part of who we are and how we experience and live in the world. To discuss the importance of sexuality is not to deny the importance of our intellectual, spiritual or emotional aspects, but to focus on an aspect of our identities, which, on some level, affects all other facets of who we are.

Carol Moog, an American psychologist and cultural critic, has conducted detailed research into sex in advertising. Her research highlights the exploitation of sexual desire and the fear of sexual inadequacy by advertisers. She writes:

Sex is rampant in advertising. And no other type of psychological imagery hits people closer to where they live. Advertisers didn't create the need for men and women to feel sexually viable, and advertisers didn't create the insecurities people have about being able to love. These are core issues in human development that cut right through to the heart of self-esteem, where people are most vulnerable. And advertisers, because they're in the business of making money, have long dangled the lure of enhanced sexuality to motivate consumers to buy.

FROM: Carol Moog PhD. 1990. 'Are they selling her lips?' *Advertising and identity*, p. 143. New York: William Morrow.

We will explore Moog's claims in more detail in the next section.

 Freud and psychoanalysis: a brief guide

The idea that sexual desire is basic to human existence was initiated primarily by the psychoanalyst and philosopher Sigmund Freud (1856–1939), whose work spanned the last years of the nineteenth century and the early part of the twentieth century. We are so used to Freudian concepts that it is difficult to imagine pre-Freudian thinking. However, it is also difficult to summarize the ideas of Freud's groundbreaking work or to do credit to the complex development of his ideas by other writers. (In fact, to do so would mean that we would have to summarize the entire tradition of psychology and psychoanalysis.) Here, therefore, are three short arguments deriving from Freud and one of his successors, Jacques Lacan.

Freud suggests that the human 'mind' (for which we can read the 'self') is divided into three parts. The first is the ego. When we think of our own identity ('Who am I?') the answers that we come up with will be a description of our ego. The ego is the 'I' that we struggle with every day of our lives. The second part of our identities is the id. The id is the part of our mind that desires instant gratification, free from any restraints. Below the level of the conscious mind, the id is a part of ourselves that simply wants to have all desires (particularly sexual desires) fulfilled. The third part of the mind, the superego, keeps the id in check. If we all acted purely on the basis of our ids our lives and society would, according to Freud, be chaotic. The superego often takes the form of morality or religious conscience.

ARGUMENT 1

Freud distinguishes between the conscious mind (the part of the mind that we are aware of, that we think with) and the subconscious mind. Many of our mental processes, some of which are linked to our sexual desires, occur without our awareness. The subconscious is a complex network of desires formed in the course of our growing up.

ARGUMENT 2

Freud's concerns play an important role in the debate about cultural difference and desire. These stress an important question: Are desires natural or are they constructed through the practices of communities? Where do desires come from? Many psychologists believe that desires are both innate and constructed through the practices of our everyday life. Two of those practices are advertising and the media. The media seem to influence sexual appetites and preferences. For example, the body types that are shown in the media in the Western world tend to be young, thin and fit. The exposure given to a certain type of body makes it more appealing. In this way the media create trends in which these bodies are more popular than other kinds. Other sexual desires are also subject to manipulation by advertising and the media.

ARGUMENT 3

The legacy of Freud has left us with an academic tradition that emphasizes the importance of sexuality (and its relation to the conscious mind). Many critics use Freud's ideas when they are analysing textual representations. We cannot ignore bodies and sex when we examine the way that texts are constructed and the effects they have in the world.

TIME TO WRITE

LOOK AT THIS cartoon. Then read the argument that follows it and write down your reasons for agreeing or disagreeing with it.

What is for sale here?

ARGUMENT

Men desire women.
Men desire the women lying on the motor car.
Therefore, men desire the motor car.
This is because men think: 'If I get the car, I'll get the girls.'
Men, therefore, are really very gullible.

Obviously desire works in more complicated ways than this. Presumably men do not believe that the women on the car are available to them. However, the selling strategy is based on the sexual desirability of the women. At some level, largely a subconscious one, men relate the car on sale to their desires. This suggests that the sexuality of both the men and the women is being used to make the car appealing. The presence of the women shifts the car into the space of sexual desire. This shift in desire, from the women to the product, is common in advertising.

Let's continue our investigation with an analysis of an advert from *Ebony* (October 1998). We are going to see how different paradigms of interpretation produce different interpretations. It is possible, for instance, to look at the way a text is constructed by breaking it up into the various elements that comprise it. This kind of interpretation describes how it is put together – a description of its construction. We referred to this as formalism and mentioned both the Russian Formalists and the New Critics in this regard. A formalist analysis of an advert would consider its formal aspects, but might not examine the ideas the text conveys (such as the moral questions it raises) in a critical way. To do this, we might have to draw on other approaches to interpretation.

ADVERTISING AND ETHICS: THE BENETTON AFFAIR

In Chapter one, we suggested that purely formalist analyses avoid contextual information and therefore do not emphasize the social and moral implications of a text. We are led to question whether this is a valid approach when we consider the ethical implications of advertising campaigns such as that launched by Benetton in the mid-1980s.

The Benetton clothing company has courted much controversy in its advertising. Images of people from diverse cultural backgrounds dressed in colourful clothing were replaced in the mid-1980s by the 'United Colours of Benetton' campaign, designed by Oliviero Toscani. The first advert to enter public discourse and become a matter for debate was for Benetton's Jesus Jeans. A woman's bottom in blue jeans is accompanied by the text, 'the one who loves me will follow me'. Protests were launched by many church groups, the most public by the Catholic Church in Italy. The Jesus Jeans advert was followed by the now-famous image of a black woman breast-feeding a white baby. This was a departure for Benetton since no products are incorporated into the advert and the connection between their range of clothing and the image is open to various interpretations. The most obvious, of course, concerns the use of 'colour' to refer to race, a move that linked the company with universal harmony and tolerance. The innovative use of bodies in the advert leads consumers to connect the name 'Benetton' with an ethical stance and set of priorities, presumably mirroring the values of a 'politically correct' youth who are interested in overturning the prejudices of earlier generations.

But the real controversy was still in store. In the early nineties, Toscani unveiled the first of his 'catastrophe advertisements'. A 1992 advert comprises a photograph of a family gathered around the bed of a newly dead AIDS victim. In the image the distraught father embraces his dead son, whose gaunt cheeks and staring eyes reveal an agonizing and tragic death. In the bottom right-hand corner, against a green background, is the slogan 'United Colours of Benetton'. This image, banned in many countries, provoked heated debate.

Because of its use of a dead 'body' in an advertising context, the representation raises many questions. Is it ethical to link a tragic death to a company's trade-name? Can adverts be used as forms of social activism, drawing people's attention to essential issues in the life of our society? Is this advert, on some basic level, more honest than those that fool us into believing that sex is about desires that are safe to enact?

Toscani, unperturbed, even inspired, by the furore, released adverts depicting an oil-sodden bird seated in a polluted river, a burning car resulting from a Mafia-related bombing in Sicily, a montage of women's and men's genitals (some of which, it was obvious, were children's penises and vaginas), and an image of dozens of multi-coloured condoms. What many saw as the last straw, though, was an advert depicting the blood-stained clothes of a (dead) Bosnian soldier laid out against a white background. How, people asked, could an image we would associate with news coverage find its way into a clothing advert? What was it doing there and is it moral to show these traces of a real death? Would a formalist analysis of this advert be adequate? This advert was withdrawn in many countries, including South Africa, following pressure from human rights groups and government. However, it is still a cornerstone in debates about advertising and ethics.

Feminist literary theory, as opposed to formalism, would allow us to examine, among other things, the way that women's bodies and sexuality are represented in a text. A feminist critic could focus on the ethical and ideological implications of a particular representation of a woman or women and evaluate the possible effects of the representation. Given that our concern here is to understand how adverts are constructed and to look at the ways that they present and use sexuality, both formalist and feminist theories are relevant. In fact, a formalist analysis is a useful departure point for a feminist analysis: we need to have a clear idea of how a text is constructed before we can interpret and evaluate the representation of women and women's bodies.

Now look at the advertisement for Opium perfume.

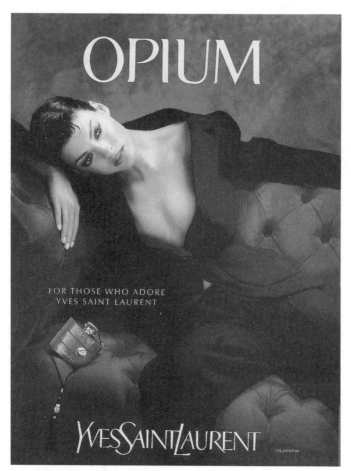

As seen in *Ebony* South Africa magazine.
Reproduced with permission from African Sales.

Let's begin by thinking about this advert in formalist terms.

◯ A formalist analysis

The advert is for a women's perfume. The image features a woman lying seductively on a leather couch. The slogan, 'For those who adore Yves Saint Laurent', explains the positioning of the woman. The slogan implies that the woman adores the (male) fashion designer Yves Saint Laurent and is therefore likely to buy the perfume. The position of her body is significant because it reinforces the message of the slogan. She stares out of the page with heavy-lidded eyes; her alluring eyes are one of the focal points of the image. Her gaze is seductive. This effect is created, in part, by the lighting from the top left, which makes her lips and eyes shine. She looks decadent, wistful and sexually available. Her top is partially loosened, revealing her bare chest and a hint of her breasts beneath. This is another focal point of the image. It implies that she is undressing or getting ready for lovemaking. The 'adoration' in the slogan is clearly sexual. Although she seems unaware of the gaze of the reader, her body is open to the male observer or reader of the advertisement.

By mentioning Yves Saint Laurent, the slogan relates the advert to a tradition of male fashion designers whose target markets (the people who advertisers hope will use the products they design) are mainly women. These men acquire cult status among women who use their products. This complicates the connotations of the word 'adore'. It implies that the woman in the image 'loves' Yves Saint Laurent and the fashion products he has created. At the same time, the position of the woman's body implies that there is a sexual element to her 'adoration'. Likewise, by wearing the perfume, the advert suggests, the woman herself will become sexually desirable.

Feminist critics, however, would examine the advert with a different set of priorities from formalists. It is possible that they would object to the stereotypical representation of women. Let's think about the advert in feminist terms.

A feminist analysis

This advert represents a woman as an object of male desire. The model has been chosen because her looks conform to a particular stereotype of female sexual attractiveness. She is young, seductively made up, dressed in black clothes and is extremely slender. The composition of the image emphasizes her large shining eyes, her parted lips and her almost-exposed breasts. These are some of the parts of women's bodies that are often highlighted in paintings and photographs. In this way the advert constructs the woman entirely in terms of a desirous male gaze. The way the woman's body is highlighted by the lighting and composition divides it into a series of sexual parts. The woman's prone position makes her look like a passive object to be consumed by the viewer's gaze. She seems to be offering herself to an invisible male viewer. This passivity is emphasized by her own heavy-lidded stare, limp wrist and parted lips, all of which suggest that the woman is drugged. This links to the name of the perfume, 'Opium', and increases the viewer's sense of her sexual passivity. Her own gaze is indirect (she does not make eye contact with the viewer) and she seems to be captured in a private moment. The viewer is thus positioned as a voyeur or 'peeping Tom'.

The written part of the advert reinforces the patriarchal stereotypes that have been set up by the image. The idea that women can 'adore' a fashion designer, combined with the position of the woman's body, suggests that women should succumb to the whims of male fashion figures. At the same time, the implication is that women need men, such as the designer Yves Saint Laurent, to help them become sexy. Moreover, in order to be attractive and sexy, women need the help of certain products, which are largely created by male fashion designers. The image defines femininity in a particular way. By linking femininity with sexual passivity and availability, the advert implies that all women wait for sexual fulfilment to come to them instead of initiating it actively. Taken as a whole, the advert links women's identity with sexuality and desirability.

This advert, then, is an example of the exploitation of feminine sexuality in advertising. In representing a sexual stereotype, the advert perpetuates the patriarchal ideal of femininity as passive and vulnerable.

✪ Does a feminist interpretation rely on a formalist analysis?
✪ Which of the two interpretations seems more appropriate and interesting? Why?
✪ What other approaches might one take to an advert such as this?

FOUR

GAZING AT
WOMEN

IN THE PREVIOUS discussion, we began to explore some issues concerning the representation of women and the uses of desire and sexuality in advertising. We also established that advertising both reflects and to some extent creates our culture. We have seen that women's bodies are often represented as objects of male desire and that sexual desire is used to sell products. Of course, this sort of representation is not limited to women, but it tends to predominate. This raises several important questions. To begin thinking critically about advertising, we need to consider differences in the ways that men and women are represented.

◯ Type-hype in advertising

Advertising relies on stereotypes of both men and women. Stereotypes distort, exaggerate or mock the actual characteristics that make up an individual. There is usually very little similarity between stereotypes and the way people view themselves. Nevertheless, comparing stereotypes gives us an idea of the ways in which gender difference is reflected and used in adverts. Look at the types we give here. See whether you can identify adverts (television or magazine) in which they are used and consider what they suggest about the (constructed) differences between men and women. You might also consider the extent to which gender representations in adverts assume a heterosexual market, or, when they don't, what stereotypes of gay men and women they advance.

TYPES OF ADVERT MEN

✪ The **solitary man** is alone on the road, has no home or emotional ties and loves physical trials in exotic places.

✪ The **family man** is a middle-class head of the household and fills the domestic roles of provider, husband and father.

✪ The **man's man** is essentially macho (either gay or straight), body-conscious and tends to be involved in 'manly' activities (sports, arm wrestling, truck driving, riding the range or beer-swilling).

✪ The **playful young man** is healthy, clean-cut and tends to play volleyball on the beach or have an exuberant time with his male and female friends in the snow, in city streets or on grassy playing fields.

✪ The **proud worker** is sweat-covered and serves society through his labours. He probably works in a factory, mine or in an emergency service. He is represented as 'the salt of the earth', a man whose opinion we can trust because he is just a straightforward guy.

✪ The **playboy** wears tailored suits and well-dressed women find his charms simply irresistible. He has an almost effortless dominance over both his world and the women he encounters.

✪ The **gent** has a city-based job, is probably speaking on a cell phone and tends to make important business deals in a loud and decisive voice, usually while looking at his watch.

✪ The **yuppie** is well-rounded in that he has an important job, is loving and sensitive, eats healthily and trains in the gym. He is independent and does not express his masculinity by dominating women.

✪ The **hunk** is a homoerotic icon with a perfect torso.

✪ The **wimp** is a nerd with glasses and a generally forlorn demeanour.

TYPES OF ADVERT WOMEN

✪ The **sporty woman** is generally in an active context, but is nevertheless impeccably clean, does not perspire and is perfectly made-up.

✪ The **good mother and wife** is defined by her ability to care for the practical needs of her children and family. She is the perfect mother, wife and homemaker.

✪ The **fox** is sexually alluring, usually has parted lips and stands provocatively, staring straight at the camera. She is not the sort of girl you imagine taking home to your mum. Her femininity is defined by her erotic allure.

✪ The **sophisticate** is immaculately groomed, perfectly dressed and poised, and is seen in social contexts linked with wealth and elegance. She tends to have men paying her attention, but remains aloof though polite.

✪ The **career woman** is shown in working environments where she is decisive, but kind. She is clearly the equal of, if not better than, her male colleagues. She wears business suits, generally with short skirts that reveal perfect legs.

✪ The **outdoors girl** is probably in the mountains or the desert as a fashion accessory for a man with a 4X4. She is radiant, affectionate and filled with a love of life and her man. Despite hiking, camping and climbing, she is never ruffled or grubby.

✪ The **'mature' woman** is often someone's grandmother and has the wisdom of age. She is gentle, kind and supportive and is either patting her ageing husband's hand or is taking care of a cherished grandchild.

✪ The **loving partner** is one of the most common types in representations of women. She embodies affection, devotion and loyalty, fulfilling the role of caretaker as well as being a gently erotic lover.

On the basis of these types, consider the following questions:

- ✪ In what ways are advert-women's roles significantly different from those of advert-men?
- ✪ How are women's bodies used in adverts as opposed to men's bodies?
- ✪ What do these differences tell us about the ideology (turn back to page 98 if you are in any doubt about this concept) of gender in our society?

Advertising is one form of social text; fashion is another. Both can be interpreted and evaluated in terms of their composition and effects. Although the discussion below may seem at first to be unrelated to advertising, the different ways in which men and women are conditioned into forms of sexuality are closely related to the way that fashion encourages people to relate to their bodies. Understanding this will leave us in a better position to approach social texts in general and adverts, which rely on these differences, in particular.

  ## Tight jeans: a philosopher speaks out

The Italian literary and cultural theorist Umberto Eco tells an interesting story in an essay 'Reading Things', included in a book about the media and culture, called *Travels in Hyperreality* (1986). He tells how he was forced to stop drinking alcohol and found himself losing weight. For the first time in years, he bought a pair of denim jeans. The sensation of wearing jeans for the first time in a while made him aware of his body and his clothing in a way that surprised him:

> The jeans didn't pinch, but they made their presence felt. Elastic though they were, I sensed a kind of sheath around the lower half of my body. Even if I had wished, I couldn't turn or wriggle my belly inside my pants; if anything I had to turn and wriggle it together with my pants ... As a result, I lived in the knowledge that I had jeans on, whereas nor-

mally we live forgetting that we're wearing undershorts or trousers. I lived for my jeans, and as a result I assumed the exterior behaviour of one who wears jeans. In any case I had assumed a demeanour. (1986:192)

The Oxford Advanced Learner's Dictionary defines 'demeanour' like this:

> **demeanour** (*US* **-nor**) /dɪˈmiːnə(r)/ *n* [U] (*fml*) a way of behaving; conduct: *maintain a professional demeanour.*

Eco then goes on to think about the implications of his experience. He considers the fact that close-fitting clothing makes the body apparent to the wearer and to the world. Let's stop for a moment. Imagine yourself wearing a large loose robe (such as a dressing gown). The garment does not match the shape of your body, which is not confined within it. If you walked, you would walk inside the robe: it would not be clear from the outside exactly how your body is moving, nor would the exact shape of your body be obvious. Now imagine wearing tight jeans. The entire experience of walking would be different. The jeans match the shape of your body and make the ways that it moves quite clear to those who observe you.

As the essay proceeds, Eco discusses the difference between the conventional Western dress of men and women and deduces that his experience of wearing jeans is very familiar to women.

[F]or women experiences of this kind are familiar because all their garments are conceived to impose a demeanour –

high heels, girdles, brassieres, pantyhose, tight sweaters. (1986: 192)

Eco makes the debatable but interesting point that the clothes we wear affect the ways that we think and, ultimately, our identity. This is, according to Eco, because certain forms of tight clothing make us conscious of our bodies and, because of this consciousness, make us 'live towards the exterior world' (1986: 193). By focusing our attention on a person's demeanour, tight clothing causes him or her to live with reference to the world, to live as though he or she is being watched or to live with a sense of performing the body.

Eco continues to focus on women's fashion:

> Women have been enslaved by fashion not only because, in obliging her to be attractive, to maintain an ethereal demeanour, to be pretty and stimulating, it made her a sex object; she has been enslaved chiefly because the clothing counselled for her forced her psychologically to live for the exterior. (1986: 194)

This idea has been echoed by many feminist and gender theorists: women's bodies are part of systems of meanings in the world that subject women to patriarchal power. The dominance of the male gaze forces women to live as though they are watched, both by men and other women. Because of the fashions that are considered to define women, attention is taken away from their intellects. Rather than living for themselves or living a life of intellectual reflection as men are supposed to do, according to patriarchal ideology, women are forced into a continual performance of their bodies and of their sexuality. While men's worlds are constructed in a way that allows them to focus on achievements of the mind that give them power and status, women are continually obliged to perform themselves as though they are objects of male attention, of a male gaze. Men act; women are looked at. Eco concludes by marvelling at those women who have overcome these obstacles, who, despite the odds being stacked against them, have managed to excel intellectually.

LOVE, SEX, BODY

TIME TO REFLECT

✪ Do you agree that the clothing we wear might influence the way we think? Why, or why not?

✪ Given Eco's argument, why do you think that monks, nuns or members of religious orders wear loose clothing?

✪ In what ways do current fashions emphasize women's bodies more than men's?

✪ How are the differences Eco discusses reflected in adverts that represent men's and women's bodies?

Considering the male gaze: another perspective

The following extract, which discusses women and the gaze, is taken from John Berger's *Ways of Seeing* (1972). Read the extract and relate it to the points made by Eco in his argument outlined above.

> To be born a woman has been to be born, within an allotted and confined space, into the keeping of men. The social presence of women has developed as a result of their ingenuity in living under such tutelage within such a limited space. But this has been at the cost of a woman's self being split into two. A woman must continually watch herself. She is almost continually accompanied by her own image of herself. While she is walking across a room or whilst she is weeping at the death of her father, she can scarcely avoid envisaging herself walking or weeping. From earliest childhood she has been taught and persuaded to survey herself continually.
>
> And so she comes to consider the surveyor and the surveyed within her as two constituent yet always distinct elements of her identity as a woman.
>
> She has to survey everything she is and everything she does because how she appears to others, and ultimately

how she appears to men, is of crucial importance for what is normally thought of as the success of her life. Her own sense of being in herself is supplanted by a sense of being appreciated as herself by another.

Men survey women before treating them. Consequently, how a woman appears to a man can determine how she will be treated. To acquire some control over this process, women must contain and interiorize it ...

One might simplify this by saying: men act and women appear. Men look at women. Women watch themselves being looked at. This determines not only most relations between men and women but also the relation of women to themselves. The surveyor of woman in herself is male: the surveyed female. Thus she turns herself into an object – and most particularly an object of vision: a sight. (1972: 46–47)

Berger argues that women experience being 'looked at' differently from men. Women are forced into a form of exteriority, of living with a sense of being looked at, and consequently, of looking at themselves. In advertising, women are subjected to the male gaze and are presented in terms of the male gaze. According to Berger, women have come to see themselves and other women through the male gaze. Women are then reduced to being bodies to look at – objects of male desire. If this is not the case for individual women, it is because they have actively resisted this tendency.

○ Applying the idea of the male gaze

Study the two adverts on this page, keeping in mind the discussion of the male gaze and the differences between male and female sexuality.

QC Classic Footwear advert. As seen in *Directions* magazine, October 1998. Reproduced with permission of John McIldowie and Associates.

Sting Occhiali advert. As seen in *Directions* magazine, October 1998. Reproduced with permission of Ken Payne Opticals.

Now answer the following questions:

✪ The women models are photographed wholly or partly naked. What is the relation, if any, between their nakedness and the products being advertised?

- How are their bodies arranged to accommodate the gaze of an observer?
- Both adverts are from men's magazine. Is the nakedness in these adverts significantly different from adverts in women's magazines?
- Do the women's bodies in these adverts have anything in common with one another?
- How do you, as the surveyor of each advert, respond to the images of the women? How is your response related to your gender?

We will be returning to these adverts soon.

FIVE

ANXIETY, ENVY, GLAMOUR AND SEXUALITY

WE ARE GOING to look at four concepts that are related to the use of bodies in adverts: anxiety, envy, glamour and sexuality.

 ## Anxiety

Look at the following extract by the feminist critic Rosalind Coward, quoted in *An introduction to women's studies* (1996).

> Advertising in this society builds precisely on the creation of an anxiety to the effect that, unless we measure up, we will not be loved. [Women] are set to work on an ever-increasing number of areas of the body, labouring to perfect and eroticize an ever-increasing number of erotogenic zones. Every minute region of the body is now exposed to the scrutiny of the ideal. Mouth, hair, eyes, eyelashes, nails, fingers, hands, skin, teeth, lips, cheeks, shoulders, arms, legs, feet – all these and many more have become areas requiring work. (1996:210)

Coward argues that the media create and then direct women's anxieties about their appearance and sexuality. The increasing number of areas 'requiring attention' is directly linked to the creation of new markets. In order to cause consumers to feel the 'need' for a given product, an anxiety needs to be created or exploited and directed towards the product that is marketed as a solution. This use or manipulation of specific anxieties leads to a fragmentation of women's bodies in advertising images. There is no holistic approach to the body: rather it comprises separate parts, each of which needs to be perfected. Women in advertisements are seldom whole or presented with equal attention being given to all parts of their bodies. They are usually reduced to being eyes, lips, hands, breasts, hair and/or legs.

Let's return to the adverts for QC Classic Footwear and Sting Occhiali sunglasses we have just looked at.

✪ To what extent do they draw attention to specific parts of women's bodies?
✪ On what particular anxieties might they rely in creating markets for products?
✪ Do you think that similar anxieties are created for men in advertising? Are there any differences in the type and degree of anxiety induced?

CONDOMS: ANXIETY AND PROTECTION

Some perceptions of bodies, and the advertising that promotes them, are life-and-death issues. Condoms present the most interesting case of a product advertised through channelled (albeit justified) anxiety. From the time when condoms began to be advertised (Durex has been advertising in the United States and the United Kingdom since the early 1960s) until 1985, they were marketed solely as a form of birth control — this despite being considered an inadequate substitute for oral contraception. Because of rapidly escalating rates of venereal disease infection, the British and American health authorities granted permission to advertisers to promote condom use. However, in spite of this, condom advertising remained short-sighted until the widespread awareness of genital herpes and HIV/AIDS in the mid-1980s. Since the late '80s, advertisers in the UK and slowly elsewhere have thought of various strategies to both eroticize the condom and to find ways of emphasizing personal safety in adverts, without lapsing into ineffectual, dull preaching.

Condom advertising is aimed at gay and straight com-munities and has attempted to both 'normalize' and glamorize condom use. In other words, adverts emphasize that they should be an inevitable and customary part of our sexual lives and that they should not be secret or a source of embarrassment. At the same time, condoms are linked with enviable lifestyles and behaviour. AIDS has remade our bodies in its own image. None of us can think about bodies and our sexuality in the same way as we did before its onset. It is this anxiety about AIDS that condom advertising both reflects and creates through a changed representation of sexuality. Overall, condom adverts do more than sell: they respond to and contribute to the sexual politics of our society.

Envy

We mentioned earlier that adverts function partly in the space between the real worlds in which we live and the worlds we desire. In this light, adverts are effective because we envy others who seem happier, wealthier or more beautiful. Adverts set up certain ways of being, living and seeming as ideals that are worthy of envy and to which we should aspire. They do the same with bodies.

Certain body types are presented as ideal. If the world of advertising is to be believed, thin and beautiful people are happier, more successful and more loveable. We are conditioned to envy certain bodies (and all that seems to follow from possessing them). Of course, only a small minority in the world has bodies like those of models. It is the envy that we experience in the face of 'perfect' bodies that entices us to purchase products that are linked to those bodies.

✪ In what ways do the adverts for QC Classic Footwear and Sting Occhiali sunglasses present bodies and lifestyles that we might consider enviable?
✪ Do you think such envy is natural, or is it created by particular representations in the media?

Glamour

Envy relates to glamour. Berger argues that 'glamour cannot exist without personal social envy being a common and widespread emotion' (1972: 148). Glamour is rooted in a society that links certain lifestyles and appearances with success. The very existence of glamour implies an unequal society, where people envy things and opportunities to which they do not have access (usually because of a lack of wealth). Advertising, in its emphasis on glamour, relies on inequality. It depends on the existence of a society in which many people want what they cannot ever have, because of their economic position, age or appearance.

Certain bodies are associated with glamour, while others are made to seem faintly embarrassing. Bodies that meet the socially constructed requirements for 'perfection' need a great deal of time and financial investment. The glamorous bodies of advertising are beyond the means of most people, as are the leisure activities in which those bodies are most often displayed.

✪ Do the adverts for QC Classic Footwear and Sting Occhiali sunglasses present glamorous people and lifestyles?

✪ In what ways is glamour apparent in the adverts and how is it used to entice you to buy the products advertised?

Sexuality

Throughout this chapter, we have emphasized the manipulation of desire in advertising. We have seen how women's bodies are presented as objects of the male gaze and we have considered some of the differences between the uses of men's and women's bodies. The use of sexual desire is central to the meaning of many adverts. Just as we have questioned the representation of bodies in advertising, we need to question the representation of sexuality. Sexuality is readily connected to certain types of bodies and to certain lifestyles. It is as if sex is the province of the young, the thin, the wealthy and the successful. So, while manipulating our sexuality, advertising also creates the sense that sexuality exists only (or most appropriately) in the forms it represents.

There are many important issues concerning the use and creation of sexuality in advertising. Think about the following two accounts of some of these issues:

ARGUMENT 1	Adverts emphasize certain sexual values above others. Not all cultures share the attitudes to sex that are assumed in advertising. Advertising, because it appeals to those with most buying power, neglects those communities that do not comprise the mainstream of society. There is thus a neglect of cultural differences in attitudes to sex and sexual relations.
ARGUMENT 2	Adverts, because they are designed to appeal to a majority of people, fail to reflect differences in sexuality. Thus, heterosexuality is the norm in the world of advertising and this contributes to the marginalization (perhaps even the suppression) of the gay and lesbian communities.

The use of bodies in advertising is only one of the ways that bodies are used in our society. We have suggested that we cannot take bodies for granted. There is no natural ideal body, for instance. Rather, the ideal is created through the types of bodies selected for representation and the ways that they are represented. The body is not simply a natural thing that always has the same meanings and significances. The meanings that are given to bodies and the values that are attached to different types of bodies (male, female, black, white, thin, fat, gay or straight) are constructed through the ways that bodies are represented in our society. Such representations exist in many forms of text and these texts need to be examined critically. In order to interpret them, we need to uncover the socially constructed codes of meaning on which they rely.

In this chapter we have focused on bodies in advertising and have looked at the use of women's bodies in print advertisements in particular. Similar questions can be asked about the ways that men's bodies are reflected and defined, just as we could ask many more questions about the relation of race and culture to the representation of bodies. Thinking about bodies has become very important in the study of cultural texts and representation. While bodies were often overlooked in analysis of texts, most theorists now think of them as crucial cultural constructions. The mass media are, unquestionably, the major sites in which this construction occurs. Therefore, in looking at any aspect of the media, we should critically look at their representation of bodies and what the cultural and sociological implications of these representations are.

4 Love, sex and poetry

LOVE, SEX AND the body form a cluster of interrelated themes, which have featured prominently in writing for many centuries. These texts cover all kinds of media, and range from popular songs through to films, TV programmes and literary writings as well. In this chapter we are going to look at some of the ways readers may usefully respond to love poetry. This involves interpreting the texts (deciding what they mean) as well as scrutinizing 'how they work'. We will be looking at the 'nuts and bolts' of how poetry makes meaning, and, in the process, how effective poetic communication is.

How do we write about love and desire? What do we write about them? In this chapter we will focus on two aspects of the discourse of love and desire. Firstly, we will look at how love poetry creates images of the 'lover'. Secondly, we will look at the ways that love poetry tries to persuade the addressee to do something or to feel in a certain way. These two aspects of love poetry sometimes come together in flattery. Flattery is a common element of love talk. A person who is attracted to someone may exaggerate the physical and mental charms of the beloved and might say, for example, that the beloved is the most beautiful or sexy person he or she has ever seen. In this way, a 'picture' of the beloved is 'painted' in words. This picture may have little resemblance to the original person. This chapter looks at some of the techniques of 'picture-making' in poetry and the way they frame people as objects of desire, from whom lovers want a certain response.

LOVE HAS MANY different faces, and before we begin reading love poetry, we are going to think about two kinds of love. First read the following summaries of the Greek myths of Eros and Psyche and of Narcissus. These myths are about two very different forms of love.

ONE

THIS IS HOW LOVE (EROS) CAME TO THE SOUL (PSYCHE)

Psyche was the daughter of an unknown king. She was so extraordinarily beautiful that men fell desperately in love with her and worshipped her as a goddess instead of courting her as a woman. The goddess of love, Aphrodite, became jealous of Psyche's beauty and sent her messenger, Eros, to make Psyche fall in love with an unworthy man.

At the same time, an oracle predicted that Psyche would wed a horrible monster on the top of a mountain. Psyche was taken outside and then carried by the wind to a castle, where Eros found her. But instead of obeying Aphrodite, Eros fell in love with Psyche and visited her every night without fail, while never allowing Psyche to see him.

Psyche consulted her two sisters about the identity of her mysterious lover. Out of jealousy, they told her. Eros then deserted her, and when their love was discovered, Psyche suffered the wrath of Aphrodite, who mistreated her

ECHO AND NARCISSUS

Narcissus was thought to be one of the most handsome young men alive at the time. He was so beautiful that people thought he must have divine parents. When Narcissus was born, the seer Tiresias was asked whether this child would live a long life and the seer replied: 'If he never knows himself.'

The nymph Echo fell in love with Narcissus, but he did not return her love. In despair she disappeared from the woods and mountains and faded away. All that remained of Echo was her voice, doomed to repeat the last words of anyone who spoke.

Narcissus had rejected the affections of many other nymphs and youths. One of them, seeking revenge for his lack of feeling, prayed to heaven: 'So may he himself love, and not gain the thing he loves.'

When the older Narcissus, discovered his image reflected in a pool, he fell in love with his own beauty. He could not reach the image in the pool and became and made her do several difficult tasks, all of which she carried out cheerfully.

However, after several further complications, the lovers could reunite as eternal partners. Psyche was finally reconciled with Aphrodite and became a goddess.

very frustrated. He could not tear himself away from the reflection and he died of sorrow by the same pool. It is said that Narcissus still keeps gazing at his image in the waters of the river Styx in the Underworld.

All the nymphs grieved for him, including Echo, and when they prepared his funeral pile, they could not find his body. In its place they found the flower, which today bears his name. (The narcissus family includes the daffodil.)

These myths are well-known examples of two different versions of love. (In fact, the word 'erotic', meaning 'used or intended to cause sexual desire', comes from Eros, the Greek god of love, and the word 'narcissism', meaning 'self-centredness', comes from Narcissus.) The tale of Eros and Psyche (who represent love and the human soul) is a story of what we might call 'ideal' or 'romantic' love, ending with the union of the two lovers. The story of Narcissus, on the other hand, reveals a character who is so much in love with his own image that he cannot establish a relationship with another person. As a result, Echo, who truly loves him, is silenced, doomed forever to repeat the last words of others. Eros and Psyche portray love as the love of the other, while Narcissus defines love as self-absorption and self-interest.

Most representations of human love lie somewhere between the two extremes of ideal union between lovers and complete self-centredness. Contrary to some ideas about love, which depict it as selfless giving without expecting anything in return, a lover usually wants something from the person to whom he or she is attracted. As a result, the discourse of love is often entwined with the language of persuasion. Some love poems are openly manipulative as they try to persuade their addressees to change their minds or to do things against their will. As we work

towards interpreting love poetry, we will keep these two aspects
– love poetry as both a portrait of the 'beloved' and as persuasive
discourse – in mind.

. . . .

WE BEGIN OUR study of love poetry with an ancient poem
depicting love as the union of two lovers. Drawn from the Bible
(and therefore translated into English), it takes the form of a dialogue between a man and a woman (presumably in Israel before
the time of Christ). First the woman speaks. She addresses the
'women of Jerusalem' and tells them about her beloved. Then the
man speaks. He talks directly to the woman, telling her what he
finds attractive about her.

TWO

LOVE AS UNION

BETWEEN LOVERS

FROM: Song of Solomon (5:10–7:9)

My beloved is all radiant and ruddy,
     distinguished among ten thousand.
His head is the finest gold;
     his locks are wavy,
     black as a raven.
His eyes are like doves
     beside springs of water,
bathed in milk,
     fitly set.
His cheeks are like beds of spices,
     yielding fragrance.
His lips are lilies,
     distilling liquid myrrh.

myrrh: a precious spice

His arms are rounded gold,
     set with jewels.
His body is ivory work,
     encrusted with sapphires.
His legs are alabaster columns,
     set upon bases of gold.

alabaster: a kind of marble,
renowned for its whiteness
and smoothness

His appearance is like Lebanon,
    choice as the cedars.
His speech is most sweet,
    and he is altogether desirable.
This is my beloved and this is my friend,
    O daughters of Jerusalem.

       …

… : a section of the poem has been left out here

How graceful are your feet in sandals,
    O queenly maiden!
Your rounded thighs are like jewels,
    the work of a master hand.
Your navel is a rounded bowl
    that never lacks mixed wine.
Your belly is a heap of wheat
    encircled with lilies.
Your two breasts are like two fawns,
    twins of a gazelle.
Your neck is like an ivory tower.

Heshbon, Bath-rab'bim, Damascus: places in the Middle East

Your eyes are pools in Heshbon,
    by the gate of Bath-rab'bim.
Your nose is like a tower of Lebanon,
    overlooking Damascus.

Carmel: Mount Carmel, a high mountain in Lebanon (today, in Israel)

Your head crowns you like Carmel,
    and your flowing locks are like purple;
    a king is held captive in the tresses.

How fair and pleasant you are,
    O loved one, delectable maiden!

delectable: delightful, lovely

You are stately as a palm tree,
    and your breasts are like its clusters.
I say I will climb the palm tree
    and lay hold of its branches.

Oh, may your breasts be like clusters of the vine,
    and the scent of your breath like apples,
and your kisses like the best wine
    that goes down smoothly
    gliding over lips and teeth.

The most striking feature of this poem is the way that the bodies of the lovers take centre-stage. As we explore more love poetry, we will find that the body plays an important role in the discourse of desire.

Let's look closely at some of the phrases that the woman (the first speaker) in *Song of Solomon* uses to describe the man she is attracted to. Here is a list:

- radiant and ruddy
- his head is the finest gold
- his eyes are like doves
- his cheeks are like beds of spices
- his lips are lilies
- his arms are rounded gold
- his body is ivory work
- his legs are alabaster columns
- his appearance is like Lebanon (a country north of Israel, renowned for its fertility and lush plant growth)

POET AND SPEAKER

When reading poetry and writing about it, it is important to keep in mind the difference between poets (the individuals who wrote the poems you are reading) and speakers. Speakers are the voices in the poems, and their views may not coincide with those of the poets. Poets often create characters in their poetry, and these characters are called 'speakers'. When writing about poetry, use the word 'speaker' to refer to the voice in the poem.

Looking at this list, it seems that the woman's entire address to the 'women of Jerusalem' is a description of her beloved's body. She has used similes and metaphors, comparing the man's body to various beautiful, exotic and luxurious things, such as gold, spices, lilies, ivory and alabaster. As these images pile up, there is an impression of overwhelming sensuous pleasure (that is, pleasure of the senses) in the body of the beloved. The woman focuses on certain carefully chosen parts of the man's body, namely his face, his arms and body, and his legs. The images she uses to describe his body point to the qualities she enjoys in it. Gold, doves, lilies, ivory and alabaster all have connotations of purity. Some of the other associations called to mind by the images are preciousness, aesthetic appeal and wealth.

The woman clearly savours her lover's body. Her descriptions are sensual: they focus on what she can see, hear, smell and touch. The fact that *Song of Solomon* is a book in the Bible implies that the sensuality it represents is holy. At the same time, because *Song of Solomon* is regarded as a metaphor for the relationship between God and humanity, it suggests that this relationship is both physical and spiritual. Thus, *Song of Solomon* challenges received or conventional ideas about the separation between the sensual and the sacred.

- ✪ How would you feel if you were the man being described here?
- ✪ In what ways do you think that the description of the man's body is exaggerated?
- ✪ Given the gender of the first speaker, does the sensuality of her desire seem surprising?

The female speaker has used so many similes and metaphors to describe the man's body that the reader almost feels swamped by the intensity of her desire for him. In this way the speaker creates a picture of the man's body in the mind of the reader, at the same time as drawing the reader into her emotions. This picture is of an overwhelmingly beautiful and desirable person.

Now read over the man's part of the poem again.

- ✪ Make a list of the images in the poem that relate to the woman's body.
- ✪ How would these images make you feel if you were the woman being addressed? For example, would you feel that you were being treated as an object of desire?

You probably noticed that the parts of her body he focuses on (her breasts, her breath and her kisses) are much more erogenous (sensitive to sexual stimulation) than the parts of his body that were depicted in the woman's part of the poem. Interestingly, while the woman tells someone (presumably her friends) how she feels about the man, he addresses her directly. It seems likely that it was not acceptable for women of the time to express desire in a straightforward manner, while men were more free to do so. At the end of his section of the poem, he even tells us and her what he wants:

I say I will climb the palm tree
    and lay hold of its branches.
Oh, may your breasts be like clusters of the vine,
    and the scent of your breath like apples,
and your kisses like the best wine
    that goes down smoothly
    gliding over lips and teeth.

He is comparing her to a plant that bears fruit (a palm tree or a grapevine). When he says he is going to climb the palm tree, this is not literally true. He intends to 'climb up' the woman's body (a very sensual image). The rest of the poem leaves her (and the reader) in no doubt of his intentions: he wants to have sex with her. This leads us to the theme of persuasion. Poets often use love poetry as persuasive discourse. Because love poetry often contains complimentary descriptions of the lover's body, flattery can be a common device to persuade a reluctant lover to have sex with the speaker. The man in *Song of Solomon* is probably using this approach, even though the woman obviously desires him equally. We will study the ways in which speakers of love poetry use their poems as persuasive discourse in much more detail later in this chapter when we look at other poems.

✪ Given the origins of the text, does the sensuality of its images surprise you?

✪ Do you think these images are appropriate for a holy text?

The most commonly used literary device in *Song of Solomon* is the simile. The structure of a simile is: 'X [something] is *like* Y [something else]'. Look at the poem again and see how many similes it contains. For a novice poet, similes are probably the easiest figure of speech to write. Now write a few lines yourself, using similes in the same way as *Song of Solomon* does. Describe someone you find attractive. You might find the following format useful (or you might choose to use your own):

My beloved's eyes are like ....
My beloved's hands are like ....
He or she is like a ....
He or she makes me feel like ....

● DOES this section appeal to your idea of what is 'sexy' in literature?
● HOW do his praises help to persuade her to make love with him?

Song of Solomon portrays two speakers, a man and a woman, each describing the body of the other in terms of extravagant praise. The speakers of love poems frequently use flattery in poetry to help them get the beloved's attention and persuade him or her to return the attention. We will examine the convention of flattery in love poetry later in this chapter when we deal with courtly love poetry. A feature that occurs in many texts is often called a literary *convention*.

Now we are going to explore another poem (which has also been translated into English) that presents love as the union of lovers, this time by an eleventh-century Indian poet called Bilhana. According to legend, he had a secret affair with his pupil, the beautiful daughter of a king. When this was discovered he was sentenced to death but was pardoned by the Indian goddess, Kali, when she read the fifty beautiful love poems he had written to his mistress.

The word 'trans.' at the beginning of the poem is an abbreviation of 'translated' and tells us that the poem was originally written in another language. In this case, it was written in Sanskrit, which was the official language of medieval India. 'Trans.' is followed by the name of the translator. So the information at the top of the poem reads 'translated by Barbara Stoller Miller'.

Translating poetry is a difficult task. Because poetry uses imagery, idioms and sound-effects that are unique to the original language and culture, some of these effects may be lost.

FROM: 'Fantasies of a Love-thief'
(trans. *Barbara Stoller Miller*)

Even now,
if I see her again,
her full moon face, lush new youth,
swollen breasts, passion's glow,
body burned by fire from love's arrows –
I'll quickly cool her limbs!

Even now,
if I see her again,
a lotus-eyed girl
weary from bearing her own heavy breasts –
I'll crush her in my arms
and drink her mouth like a madman,
a bee insatiably drinking a lotus!

lotus: an Indian waterlily; often associated with intoxication

insatiably: without satisfaction

Even now,
I remember her in love –
her body weak with fatigue,
swarms of curling hair
falling on pale cheeks,
trying to hide
the secret of her guilt.
Her soft arms
clung
like vines on my neck.

Even now,
I remember her:
deep eyes' glittering pupils
dancing wildly in love's vigil,
a wild goose
in our lotus bed of passion –
her face bowed low with shame
at dawn.

Even now,
if I see her again,
wide-eyed,
fevered from long parting –
I'll lock her tight in my limbs,
close my eyes, and never leave her!

fatigue: tiredness

vigil: a watch, usually at
night

Now answer the following questions in your journal:

✪ Why does every stanza begin with 'Even now'?
✪ How does the speaker feel about his beloved? Which words
and phrases in the poem tell us about his feelings for her,
and especially for her body?
✪ What does he want to do with her?
✪ Of what, if anything, is the speaker trying to convince his
beloved and his readers?

STANZAS

'Fantasies of a Love-thief' is divided into five stanzas. A stanza is a unit in a poem. We say that poems are divided into stanzas. Strictly speaking, 'stanza' refers to a group of lines sharing the same structure (e.g., the units in a villanelle) and other irregular units should be called 'verse paragraphs'. However, it is now common to use 'stanza' to refer to any group of verse lines. There is a line open between stanzas, showing that one stanza has ended and another is beginning.

This poem, like *Song of Solomon*, describes the body of the beloved in detail. There are several references to her body: to her face, her breasts, her mouth, her arms and so on. The speaker has used repetition (of the words 'Even now', 'if I see her again' and the promises at the end of the first, second and third stanzas) to make his point more strongly. We would gather from the poem that this is a forbidden love affair, even if we had not read the contextual information just before it, because the speaker mentions 'guilt' (stanza 3), 'shame' (stanza 4) and 'long parting' (stanza 5). When we read 'long parting', we realize that the lovers have been separated against their will, probably by the girl's parents. Why would they do this? A likely reason lies in the marriage conventions of India in the eleventh century. Parents arranged marriages, rather than allowing young people to choose their own partners. There is a very strong class system (called 'caste') in India and it is still not really acceptable for members of one caste to marry members of another. The daughter of a king would have been destined for marriage with the son of another royal family, and for her to be involved with a mere court poet would have been shameful.

TIME TO DISCUSS

✪ What do you think of arranged marriages as opposed to marriage by choice?
✪ Does the speaker's enduring passion, despite having been separated from the girl, persuade you of the sincerity of his affections?
✪ Does the poem contain any persuasive message? In other words, what, if anything, does the speaker want from the girl?

The images relating to the body in 'Fantasies of a Love-thief' are very sensuous. For instance, the word 'lotus' (India's most famous flower) is used several times. The seed of the lotus was supposed to lull people who ate it into a dreamy state so that they forgot all their obligations and simply enjoyed the pleasures of the body. Lotus seeds were also claimed to be addictive, so that people who ate them were in danger of forsaking all their responsibilities and becoming enslaved to the senses. The word 'lotus' highlights the forbidden but delightfully physical aspects of the love affair. It implies that the lovers spent their time together in an almost unreal frame of mind.

Here are some other sensuous images from the poem. Read the information box on denotations and connotations; then explore the denotations and connotations of the images in 'Fantasies of a Love-thief' by completing the table that follows.

DENOTATIONS AND CONNOTATIONS

Often the 'meaning' of language in poetry goes beyond its **denotation** and draws on its **connotations**.

Denotation is the thing that is 'pointed to' by a word or phrase. In the line 'O my love is like a red, red rose' (from Burns's poem, 'A red, red rose') the denotation of the phrase 'a red, red rose' is simply that: a rose that is red in colour. **Connotations** are the qualities and values associated with the thing being referred to. Words with the same or similar denotations may have very different connotations. For example, we associate passion, beauty and romance with the image of a red rose. Think about the different connotations of the following groups of words:

fat; plump; well-rounded; overweight; obese
tense; highly strung; neurotic

In each of these groups, the words have the same denotation, but very different connotations. Connotations are often connected with value judgements. For example, if you refer to someone as 'slender', this implies that you approve of that person's body shape; but if you refer to him or her as 'skinny', it implies that you do not.

IMAGE	DENOTATION	CONNOTATION
body burned by fire from love's arrows		
I'll quickly cool her limbs		
weary from bearing her own heavy breasts		
a bee insatiably drinking a lotus		
body weak with fatigue		
Her soft arms / clung / like vines on my neck		
dancing wildly in love's vigil		

By filling in the table, linking particular denotations and connotations to phrases from 'Fantasies of a Love-thief', you have explored how the speaker uses language to create a word-picture of the woman in the poem. The words and phrases that he uses enable the reader to 'see' the woman through his eyes and to value aspects of her body that he values. All the poem's readers might not have the same standards of beauty as the speaker.

PLATONIC THINKING

Plato was a Greek philosopher who lived in the period before Christ's birth. He argued, among other things, that the world of the spirit was the only realm in which perfection could be found. It follows from this idea that the realm of the senses is fundamentally flawed.

In 'Fantasies of a Love-thief', the connection between love and sexuality is very interesting. Love, in the *Platonic* tradition, is seen as separate or divorced from the body – a matter of the emotions and the spirit. The Platonic lover loves because he or she sees in the other person an image of a perfect spiritual ideal. (This is why the phrase 'Platonic relationship' is used to refer to a relationship that is not sexual.) By contrast, in 'Fantasies of a Love-thief', as in *Song of Solomon*, the speaker's devotion to the woman is seen most clearly in his tributes to her body. We have noted how carefully, indeed artistically, the speaker depicts the world of the senses, namely the body and its experiences, in particular sexuality and desire.

TIME TO DISCUSS

✪ Does this poem describe an attraction that is 'only' physical?

✪ Does the emphasis on the body degrade the speaker's affection, making it inferior to more spiritual devotion?

✪ Do you find the speaker's explicit references to sex and love-making offensive?

✪ There are also implicit references, such as 'Her body weak with fatigue' in the third stanza and 'at dawn' (that is, after a night of love-making) in the fourth stanza. Could these phrases make the poem pornographic? If you think so, explain the reasons for your response.

TOWARDS AN INTERPRETATION OF 'FANTASIES OF A LOVE-THIEF'

Use your exploration of the images of the woman's body in 'Fantasies of a Love-thief' as a starting point for an essay on the poem. You should discuss the following:

✪ What does the title, 'Fantasies of a Love-thief', mean?

✪ What are the central themes of the poem?

✪ How does the speaker use language to communicate his point/s to the reader?

✪ What does he wish for?

You might also choose to express an opinion about whether the poem is a 'good' one or not. If you do, though, you should remember the point made in *Critical conditions*, namely that our evaluation depends on our assumptions about literary criticism and interpretation.

THREE

FLATTERY AND COURTLY LOVE CONVENTIONS

AT THE BEGINNING of this chapter we said that flattery is a common feature of some love poetry. In the next section, we are going to look at four poems where flattery is prominent. We will examine different forms of poetic persuasion and see how they are related to a long tradition of discourse about love. This legacy stretches back to the Roman poet Ovid (who lived and wrote about 2 000 years ago), but is rooted more firmly in the codes of *courtly* love. This version of romantic love defined relationships between aristocratic lovers (knights, lords and ladies) from the eleventh to the fourteenth centuries – known as the Middle Ages.

Courtly love is governed by elaborate rules. A nobleman in love with a married woman of high rank had to prove his devotion by performing heroic deeds and by presenting her with anonymous amorous writings (love letters and love poems). If his suit was successful, she would return his affections, the lovers would pledge themselves to each other and the relationship would sometimes be consummated (that is, completed by sexual intercourse). At this stage, absolute faithfulness and secrecy had to be maintained, particularly because this was an extra-marital relationship.

In literature, the elements of courtly love inspired many great writers. Of these, the sonnets of the Italian poet, Petrarch (1304–1374), dealing with his unrequited passion for a beautiful young French woman, Laura, had perhaps the greatest influence on English literature. This influence is referred to as the 'Petrarchan tradition'. Typically, in these works the lover praises the beauty and purity of his beloved and pledges undying loyalty to her even though he knows that she may remain unattainable to him.

Here is an example of a classic courtly love poem written by Sir Philip Sidney

(1554–1586). It comes from a series of poems called 'Astrophel and Stella' ('stargazer' and 'star'), which recalls Petrarch's praise of Laura and his expression of frustrated love:

FROM: 'Astrophel and Stella'
Sir Philip Sidney

Who will in fairest book of Nature know
How beauty may best lodged in beauty be,
Let him but learn of love to read in thee,
Stella, those fair lines which true goodness show.
There shall he find all vices' overthrow,
Not by rude force, but sweetest sovereignty
Of reason, from whose light those night birds fly,   night birds: emblems of vice
That inward sun in thine eyes shineth so.
And, not content to be perfection's heir
Thyself, dost strive all minds that way to move,
Who mark in thee what is in thee most fair.
So while thy beauty draws the heart to love,
    As fast thy virtue bends that love to good.
    'But ah,' Desire still cries, 'give me some food.'

Typically, although the speaker praises the purity and beauty of the beloved (the distant, unattainable Stella), the poem ends on a note of frustration.

FURTHER READING

Poetry on the theme of love lost, such as 'Astrophel and Stella', is very common. Here are two other poems on this subject. These poems both describe parting after disagreements. Look at the way the attitude to the beloved changes in each poem. In both poems, the speaker struggles with the desire to both praise and blame the beloved at the same time.

SINCE THERE'S NO HELP ...
Michael Drayton

Since there's no help, come let us kiss and part:
Nay, I have done; you get no more of me,
And I am glad, yea, glad with all my heart,
That thus so cleanly I myself can free;
Shake hands for ever, cancel all our vows,
And when we meet at any time again,
Be it not seen in either of our brows
That we one jot of former love retain.
Now at the last gasp of Love's latest breath,
When, his pulse failing, Passion speechless lies,
When Faith is kneeling by his bed of death,
And Innocence is closing up his eyes,
     Now if thou wouldst, when all have given him over,
     From death to life thou mightst him still recover.

THEY FLEE FROM ME ...
Thomas Wyatt

They flee from me that sometime did me seek
With naked foot stalking in my chamber.
I have seen them gentle, tame, and meek
That now are wild, and do not remember
That sometime they put themself in danger
To take bread at my hand; and now they range,
Busily seeking with a continual change.

Thanked be fortune it hath been otherwise
Twenty times better, but once in special,
In thin array after a pleasant guise
When her loose gown from her shoulders did fall,
And she me caught in her arms long and small,
Therewithal sweetly did me kiss,
And softly said, 'Dear heart, how like you this?'

themself: themselves

in special: especially
guise: appearance

small: thin
therewithal: as well

It was no dream: I lay broad waking.   broad waking: wide awake
But all is turned thorough my gentleness   thorough: through
Into a strange fashion of forsaking,
And I have leave to go of her goodness,
And she also to use newfangleness.   newfangleness: strangeness
But since that I so kindly am served,   kindly: ironic for unkind
I would fain know what she hath deserved.   fain: very much

Now look at the following poem by Andrew Marvell (1621–1678)
and see what the speaker is trying to persuade his mistress to do.

TO HIS COY MISTRESS   mistress: beloved
Andrew Marvell

Had we but world enough, and time,   world: space
This coyness Lady were no crime.   coyness: modesty
We would sit down, and think which way
To walk, and pass our long love's day.
Thou by the Indian Ganges' side   Ganges: a river in India
Should'st rubies find: I by the tide
Of Humber would complain. I would   Humber: a river between
Love you ten years before the Flood: North and South
And you should if you please refuse   Humberside, England
Till the conversion of the Jews.
My vegetable love should grow
Vaster than empires, and more slow.
An hundred years should go to praise
Thine eyes, and on thy forehead gaze.
Two hundred to adore each breast:
But thirty thousand to the rest.
An age at least to every part,
And the last age should show your heart.
For Lady you deserve this state;
Nor would I love at lower rate.
But at my back I always hear
Time's winged chariot hurrying near:

And yonder all before us lie
Deserts of vast eternity.
Thy beauty shall no more be found;
Nor, in thy marble vault, shall sound
My echoing song: then worms shall try
That long preserved virginity:

honour: modesty

And your quaint honour turn to dust;
And into ashes all my lust.
The grave's a fine and private place,
But none I think do there embrace.
Now therefore, while the youthful hue
Sits on thy skin like morning dew,
And while thy willing soul transpires
At every pore with instant fires,
Now let us sport us while we may;

amorous: loving (here,
lustful)

And now, like amorous birds of prey,
Rather at once our time devour,

slow-chapt: slowly
consuming

Than languish in his slow-chapt power.
Let us roll all our strength, and all
Our sweetness, up into one ball:
And tear our pleasures with rough strife,

thorough: through

Thorough the iron gates of life.
Thus, though we cannot make our sun
Stand still, yet we will make him run.

'To His Coy Mistress' seems to be an elaborate and highly exaggerated attempt by the speaker to urge his 'mistress' to have sex with him. At the core of his persuasive strategy is the idea of *carpe diem* (Latin for 'seize the day'), which urges the lovers to live for the moment. The humour of the poem comes from the way it draws on various social and poetic conventions, and from the extensive use of hyperbole (exaggeration). Although essentially witty in tone, 'To His Coy Mistress' also challenges conventional moral values and addresses the more serious aspects of human mortality.

The poem is divided into three sections, each different in tone, pace and concern. Each section represents a distinct part of the speaker's developing argument and the poem as a whole forms a neat, yet logically flawed, syllogism.

In the first section of 'To His Coy Mistress' the speaker describes a leisurely life, completely removed from the restrictions of time, in which he would be able to indulge his mistress fully and to wait for her to consent to have sex with him. By contrast, the second section depicts a grim and barren eternity. The

SYLLOGISMS

The *Oxford Advanced Learner's Dictionary* defines a syllogism as a form of reasoning in which the conclusion is drawn from information given in two statements. The following is an example of a syllogism:

All men are mortal.
Socrates is a man.
Therefore, Socrates is mortal.

speaker tries to impress upon his mistress the urgency of consummating their relationship by using images of death and decay to indicate the constraints on Platonic love. In the final section, the speaker enthusiastically provides a 'solution' to the limitations imposed by inevitable mortality.

Now that we've summarized the poem, we are going to look closely at some of the details. The opening lines of 'To His Coy Mistress' immediately set out the limitations imposed on Platonic love by time and space:

> Had we but world enough, and time,
> This coyness Lady were no crime.

The speaker goes on to show that a timeless existence ('world enough, and time') is unrealistic, because time passes very swiftly. The lady's coyness itself is not a 'crime', and is playfully called one only because the speaker and his mistress do not have 'world enough, and time' in which to enjoy the slow and deliberate courtship expected by society. Cleverly, the speaker does not immediately dispute the idea of courting his mistress at leisure. An important part of his seduction technique is to convince the woman that he sincerely loves her, and is not just motivated by lust. The verbs in the following lines shift subtly from the static – 'sit' and 'think' – to gradual, unrushed action – 'walk' and 'pass', showing us that the speaker is not in a hurry to secure his mistress's consent. Similarly, the distance between the 'Indian Ganges' and the 'Humber', as well as the extension of time from 'before the Flood' to the 'conversion of the Jews' (which was supposed to be a sign that the world was coming to an end), imply that the speaker's love is endless and boundless. In an ideal world, he would be content to endure, with patience, the restraints on consummating their relationship.

The speaker's exaggerated claim that he has an unlimited capacity to adore from afar is strengthened by his imaginative and exotic extension of time and love in the following lines:

> My vegetable love would grow
> Vaster than empires and more slow.
> An hundred years should go to praise
> Thine eyes, and on thy forehead gaze.
> Two hundred to adore each breast:
> But thirty thousand to the rest.

The suggestion here is that, given infinite time, he would let his love mature slowly and become immensely large before being consummated by sexual intercourse. This is reinforced when the speaker expresses his adoration in terms of the time he is going to spend praising each part of his mistress's body. This period increases from 'an hundred years' to an 'age', which would finally reveal her 'heart'. The endless worship of every aspect of the mistress's being is intended to convey the extent of the speaker's love and his acknowledgement of her worth:

> For Lady, you deserve this state;
> Nor would I love at lower rate.

Yet, at the same time, aspects of the first section alert the reader to hidden motives in the speaker's claims of love. For example, the extravagance of his adoration parodies the conventions of courtly love and ironically undermines our sense of his sincerity. (In the tradition of courtly love poetry, the lover was 'supposed' to make exaggerated claims about his lady's beauty and his adoration. But the speaker in 'To His Coy Mistress' outdoes any such claims.) Moreover, the words 'to the rest' place emphasis on the physical and, indirectly, on the sexual aspects of love. As a result we doubt the speaker's intention to devote the 'last age' to his mistress's heart, and suspect that he is going to focus on sexuality instead.

The underlying physicality of the first section of 'To His Coy Mistress' is expressed more directly in the second part of the poem, which has, at times, an urgent tone and pace (notice how the pace of the poem speeds up). The section is dominated by a stark image of death – 'Deserts of vast eternity' – and barrenness and ugliness are contrasted with the romantic view of the previous section. Through the persuasive trickery of his words, the speaker ignores any religious idea of a heavenly eternity in order to concentrate on the horror of decay and the way death destroys youthful beauty and innocence:

> Thy beauty shall no more be found;
> Nor, in thy marble vault, shall sound
> My echoing song: then worms shall try
> That long preserved virginity.

The change in tone in these lines is accompanied by a narrowing of physical space. The emphasis is on the confinement of the grave, as opposed to the boundlessness of the previous section. At the same time, the sense of the exotic in the first part (rubies and the Ganges) is transformed into sterility (marble is cold and clinical) and the passionate into the repulsive. The suitor implies that the desolation of death is the opposite of the pleasures his love will provide. Yet, his 'scare tactics' contain elements of a more serious concern. Social conventions may preserve beauty and purity beyond their usefulness and can become absurd in the context of humankind's race against time. Furthermore, death will not honour these values and the lady's 'long preserved virginity' will be 'tried' (that is, destroyed) by worms. So while the speaker may challenge conventional moral values, he also subtly suggests that nature takes no notice of them.

However, the speaker's deliberate emphasis on the seriousness and ugliness of death masks his sexual agenda. Underlying his apparent sincerity is an ulterior motive. There is a certain delight in the speaker's play on human fears and in his description of the horrors of the grave. Moreover, unlike the sincerity we sensed in 'Fantasies of a Love-thief', we feel that the speaker of 'To His Coy Mistress' has adopted a mocking tone. Consider, for example, the ironic tone in the following lines:

> And your quaint honour turn to dust;
> And into ashes all my lust.
> The grave's a fine and private place,
> But none I think do there embrace.

The word 'quaint' also had another meaning during Marvell's time: it was often used as a slang word for female genitals. This adds a further dimension to our reading of the line!

Whether one regards 'quaint', in seventeenth-century terms, as meaning 'fastidious', or whether one interprets the word as connoting curious, old-fashioned attitudes (as we tend to do today), the speaker is mocking his mistress's 'honour' (her chastity). This dismissal of the lady's 'honour' has further resonances when seen in relation to the courtly love tradition and the Platonic perception that human beauty should reveal divine beauty, rather than act as an invitation to lust.

In the final section of the poem, the speaker offers an alternative or 'solution' to the problems produced by the limitations of time. Sexual gratification will defy time's 'slow-chapt power':

> Now, therefore, while the youthful hue
> Sits on thy skin like morning dew
> And while thy willing soul transpires
> At every pore with instant fires,
> Now let us sport us while we may ...

The comparison of the lady's youthful complexion with 'morning dew' conveys the transience and vitality of youth, the inevitable loss of which the speaker has shown in the second section of the poem. In the next two lines, the speaker suggests that the lady's coyness cannot prevent her amorous spirit ('willing soul') from showing in her flushed face. Thus, 'youthful hue' may be interpreted both as indicating the bloom of youth and sexual excitement. (Alternatively, the lady may be flushed with embarrassment at such a direct assault on her sensibilities!) The speaker hopes that his argument will transform 'coyness' into a 'willing soul', and 'long preserved virginity' into 'instant fires' of passion.

Startling images of unrestrained physicality and sexual energy replace the discourse of persuasion that is evident in the previous sections of 'To His Coy Mistress'. The speaker unmistakably suggests an immediate, forceful gratification of sexual desire and the abandonment of romantic pleasantries:

> And now, like amorous birds of prey,
> Rather at once our time devour,
> Than languish in his slow-chapt power.

The repetition of 'now' and the energy connoted by the words 'devour', 'tear' and 'rough strife' all add urgency into the argument. Time has been contracted throughout the poem until, finally, only the instant remains. At the same time, the wide expanses of the first section (which were reduced to the confines of the grave) are again constricted to the indivisibility of the sexual act represented by 'one ball'. The reference to 'the iron gates of life' is in keeping with the idea of imprisonment (not only by time and space, but also by social convention) so carefully established by the speaker. The joint pleasure of passion, by implication, is the only release from mortality.

The poem ends by irreverently alluding to the Bible. Although the lovers cannot, like Joshua, make the sun stand still (Joshua 10:12–14), they can, through the energy of their love-making, defy time – filling each day to such an extent that the sun is hard pressed to keep pace:

> Thus, though we cannot make our sun
> Stand still, yet we will make him run.

The suitor's intentions come as no surprise. They have been implied throughout the poem. The response of the mistress to whom the argument is addressed remains enigmatic as she is silent. Unlike *Song of Solomon*, this is often the case in courtly love poetry: the beloved's response remains unknown.

Like 'To His Coy Mistress', our next poem also challenges the courtly love tradition, but in a very different way. Shakespeare's 'My Mistress' Eyes' is one of the best-known love poems in English. (Does this mean it is one of the best? Read it carefully and then decide for yourself.)

● IMAGINE that the poem is addressed to you. How would you feel about the speaker's persuasive argument?

MY MISTRESS' EYES
William Shakespeare

My mistress' eyes are nothing like the sun
Coral is far more red, than her lips red,
If snow be white, why then her breasts are dun:
If hairs be wires, black wires grow on her head:
I have seen roses damasked, red and white,
But no such roses see I in her cheeks,
And in some perfumes is there more delight,
Than in the breath that from my mistress reeks.
I love to hear her speak, yet well I know,
That music hath a far more pleasing sound:
I grant I never saw a goddess go,
My mistress when she walks treads on the ground.
    And yet, by heaven I think my love as rare,
    As any she belied with false compare.

dun: grey

roses damasked: an
old variety of rose
with velvety petals

reeks: a neutral term
meaning 'smells' in
Elizabethan times

go: walk

belied: made untrue,
contradicted/compare:
comparison

Without re-reading the poem, 'My Mistress' Eyes', write down
your *first* response to it. Use the following format, which we will
return to later in our discussion.

'My Mistress' Eyes' is good / bad / mediocre (choose one).
The poem is about ....
The speaker uses .... to help him
get his meaning across.
I like / do not like the poem because ....

As we look more closely at the poem, you might modify your
first impressions and work towards a more inclusive interpretation and evaluation. Notice that what you have just written contains both interpretive and evaluative elements. When you write
'The poem is about ...', you are making an interpretive claim. But
when you write 'I like the poem because ...', you are evaluating or
judging it.

THE SONNET

A sonnet is a poem that has fourteen lines and an elaborate rhyme scheme. A *Petrarchan sonnet* is divided into an octet (a section made up of eight lines) rhyming *a b b a a b b a* and a sestet (a section comprising six lines) rhyming *c d e c d e* or *c d c d c d*. There is normally a shift in meaning or emphasis between the octet and sestet. The Shakespearean sonnet typically has three quatrains (sections made up of four lines) and a final couplet (two lines), which often casts new light on the rest of the poem . The rhyme scheme is usually *a b a b, c d c d, e f e f, g g* or *a b a b, b c b c, c d c d, e e*. 'My Mistress' Eyes' use the first rhyme scheme of the Shakespearean sonnet. Because of their fixed form, sonnets are extremely challenging to write.

When you write about sonnets, you do not necessarily have to refer to the technical aspects of this form of poetry as outlined above, but understanding the layout can be very helpful in following the argument of the poem.

Now let's look more closely at 'My Mistress' Eyes'. We are going to consider what the speaker says about his mistress' body. Fill in the following table, showing how parts of the body are described in the poem:

Part of the body	Description
eyes	'nothing like the sun'
lips	
breasts	
hair	

Shakespeare is mocking the Petrarchan convention of the *blazon*, which includes a listing of parts of the body from the hair down to the feet and uses hyperbolic similes to describe them, e.g., lips like coral, teeth like pearls etc.

As you can see, there are no descriptions in the poem that could be called 'flattering'. This poem's approach to the body of the beloved is exactly the opposite of that taken in *Song of Solomon*, where extravagant praise is used to evoke physical beauty and desire. In contrast, the speaker of 'My Mistress' Eyes' spends his time telling the reader what his mistress' body is not like. For example, he writes 'Coral is far more red, than her lips red'. We can rephrase this by saying 'her lips are not as red as coral'. What is the speaker saying here? Is he suggesting that her lips should be like coral, or that they would be coral-coloured if she were really beautiful? Is he suggesting that she is not beautiful? If so, this is a very unconventional claim to make in a love poem.

There are two possibilities here. They are:

INTERPRETATION A

The speaker is saying that his mistress' body is not so beautiful that she can be compared to precious substances as is the case with other courtly love poems. The speaker implies that the conventions of courtly love seem ridiculous because his beloved is in fact rather ordinary looking. However, he still thinks that she is rare and precious because he loves her.

INTERPRETATION B

The speaker is saying that his mistress' beauty is not the kind that can be praised extravagantly in images such as we found in *Song of Solomon*. She is beautiful, but the poetic conventions of comparing the lover's body to the sun, coral, roses and so on do not do justice to her kind of beauty. He wants to describe her inner beauty.

✪ Which of these two interpretations do you find more convincing? Why?

✪ Why did you choose the interpretation you did? Was it because you liked it, because you thought the poem supported it, or was there another reason?

As you saw in *Critical conditions*, your choice of interpretations is governed by your critical stance. You may have decided that Interpretation B is more convincing simply because you think love poetry should praise the beauty of the person it describes. Or you may have decided that Interpretation A is more fully supported by the poem. In either case, you are saying something about what you think love poetry is or should be.

Now look at the last four lines. The speaker says:

I grant I never saw a goddess go,
My mistress when she walks treads on the ground.
And yet, by heaven I think my love as rare,
As any she belied with false compare.

He begins by saying that his mistress 'treads on the ground'. In other words, she is not a 'goddess', but an ordinary human person. This summarizes the descriptions in the rest of the poem, which tend to make the reader feel that his mistress is less beautiful than the women described in other texts (such as *Song of Solomon* and 'Fantasies of a Love-thief'). The last two lines, though, reverse everything the speaker has been saying up to now. He says: 'I think my love as rare [beautiful]/As any [description] she belied [turned into a lie] with false compare [comparison]'. In other words, his mistress is as beautiful in his eyes as any other woman who has been extravagantly, but falsely, described in conventional love poetry. Here Shakespeare summons the whole tradition of flattery and praise of the beloved that characterizes love poetry. He shows that this kind of poetry exaggerates and falsifies the qualities of the beloved, making him or her unrealistically beautiful.

Here is an example of the idealized representations of women that Shakespeare is reacting against:

Her complexion is as white as snow. Her lips are two branches of coral; each cheek is crossed by a lily and a rose. In the place where we should see the eyes, we do not see white or iris; instead there are two suns which cast their rays like those one notices in flames or darts. The eyebrows are as black as ebony and are shaped like two bows ... Above is the clear brow, looking like a mirror, and above this a little Cupid is sitting upon a throne, and to embellish the work the hair waves round about in different patterns. Some hairs have been made like chains of gold, others are like nets and webs, and most hang like lines with fishhooks at the end, complete with bait to catch their prey; there is a number of hearts that are the bait to be swallowed, and, among the others, one larger than its companions that has been placed below the left cheek in such a way that it appears to be hanging from the ear of this rare beauty ...

FROM: (Leonard Foster, 1969. *The Icy Fire*, xi. Cambridge: Cambridge University Press.)

In contrast to this idealized version of feminine beauty, Shakespeare's poem describes a 'real' woman, one who has physical faults and imperfections. But he thinks she is as beautiful as the women described in conventional love poetry. To a large extent, 'My Mistress' Eyes' is about the convention of physical flattery in love poetry. It shows how easy it is to praise someone extravagantly, without seeing what he or she really looks like. These conventions create a template of the ideal body according to certain notions of beauty, but they are not adequate when describing a real, flesh-and-blood person with specific psychological needs and desires.

- ✪ What do you think of 'My Mistress' Eyes' after you have explored the images a little? Write an interpretation and evaluation of the poem.
- ✪ How does this interpretation differ from the one you wrote after you had read the poem once? Have you changed your opinion about the poem's worth or its meaning?
- ✪ How did the information about the literary context of this sonnet help you to interpret it?

It is probable that your first, subjective response formed the starting point of your interpretation and evaluation when you wrote about it on the second occasion. But in its final form, your views should go beyond subjective impressions. You should express an informed opinion, supported by evidence from the poem.

I mentioned earlier that 'My Mistress' Eyes' is one of the most famous love sonnets in the English language. It is so famous that it has been borrowed and re-written several times. In the following song by the British musician Sting (who was a high school English teacher before he became a musician) the opening line of 'My Mistress' Eyes' is borrowed:

SISTER MOON
Sting

Sister moon will be my guide
In your blue blue shadows I would hide
All good people asleep tonight
I'm all by myself in your silver light
I would gaze at your face the whole night through
I'd go out of my mind, but for you

Lying in a mother's arms
The primal root of a woman's charms
I'm a stranger to the sun
My eyes are too weak
How cold is a heart
When it's warmth that he seeks?
You watch every night, you don't care what I do
I'd go out of my mind, but for you
I'd go out of my mind, but for you

My mistress' eyes are nothing like the sun
My hunger for her explains everything I've done
To howl at the moon the whole night through
I'd go out of my mind, but for you

Sister Moon

This is a lot like the conventional love poetry that we have just looked at. Bear in mind that this is a pop song, and has been performed for thousands of people. In this light, how would you feel if you were Sting's 'mistress'? Would you be flattered that he had written and performed a song about you? Do you think the poem/song intends to make the woman feel this way? What response does it aim to produce?

Let's look at how 'Sister Moon' makes the woman to whom it is addressed feel flattered and recognized. Our first encounter with the song associates the woman with night. The title associates the moon with femininity. Stop reading now, and fill in the following table, giving the connotations you associate with night and the moon. How do these make you think about the ideas of love and desire?

PHRASE FROM 'SISTER MOON'	CONNOTATIONS	CONNECTION WITH LOVE AND DESIRE
Sister Moon	contrasts with the ideas of 'brother' and 'sun'; associations with coolness, light, difference, femininity (women are traditionally associated with the moon and men with the sun)	
blue blue shadows		
gaze at your face the whole night through		
silver light		
howl at the moon the whole night through		

The chorus of the song (repeated at the end of each verse) is 'I'd go out of my mind, but for you'. This crucial statement tells us that the speaker is desperately in love: his very sanity depends on her. Nevertheless, being in love seems to have made him slightly crazy already, as we see in 'To howl at the moon the whole night through', which reminds us of wolves or dogs howling at the moon. The line compares the speaker's feelings to those of an animal during the mating season. Sting uses an archetypal symbol of longing and madness to express his desire. This brings to mind the sexual drive we associate with love and it also makes us feel that the speaker is in the grip of powerful, almost irresistible needs. This recalls the common sense of love as a kind of madness, but one that perches on the edge of sanity, since the things that remind the speaker of the beloved keep him from going completely insane.

There are powerful images of desire in the song. Does the speaker want:

1 love?
2 a surrogate 'mother'?
3 a faithful 'mistress'?
4 sex?

Which alternative/s did you choose? Given the rest of the poem/song, I am inclined to think that 2 is the most likely, although all of them are possible. He uses images of insanity to imply that 'sister moon' is his last link to sanity, so we can gather that he wants her to keep him sane. In the second verse, he uses images of motherhood to suggest that he is in the role of a child in relation to her. Go back to the poem and examine these more closely. Try to decide which of the many connotations of motherhood are relevant here. Is it positive? Nurturing? 'Essentially' feminine? Sigmund Freud argued that wives have to take on the role of a mother to their husbands in order to ensure their faithfulness and love. Freud's view is very controversial, but it seems close to the relationship that the speaker describes in the lines:

Lying in a mother's arms
The primal root of a woman's charms

These lines position the man in a needy and dependent role, which is reinforced by the later claim that 'I'd go out of my mind, but for you'.

Before we leave 'Sister Moon', let's consider how Sting refers to Shakespeare's 'To Mistress' Eyes'. In the third verse, we read:

My mistress' eyes are nothing like the sun
My hunger for her explains everything I've done.

When you read the quotation from Shakespeare's poem, what do you think of? For me, this line brings to mind the whole of 'My Mistress' Eyes', especially the final two lines with their declaration of passionate love for the mistress. Sting goes one step further in describing the intensity of his feelings. He says his 'hunger' (which must be a reference to desire) for the woman is the reason for everything he has (ever) done. In other words, he lives for her.

TIME TO WRITE

Write some notes for an interpretation of 'Sister Moon' in response to the following questions.

✪ Does the quotation from Shakespeare's poem 'My Mistress' Eyes' strengthen or add to the song? If so, how?

✪ How do the references to night and the moon, as well as to women as mothers, enhance the speaker's claims about his feelings?

✪ What overall impression do you gain of the body of the woman being addressed in 'Sister Moon'?

Now reflect on whether you like the song or not, and what you think it is saying. Fill in the spaces in the paragraph below:

The main idea in 'Sister Moon' is ....
The speaker conveys the idea of his lover's body as ....
He writes the song in order to persuade her to ....

This song is effective / ineffective (choose one) as a love poem
because ....

Before we end this section, we need to consider an important point: all the women in these poems are seen from the perspective of male speakers. Whether the speakers are sincere or manipulative, respectful or not, the images of love, sex and body are determined by masculine values and desires. The attributes and qualities that are praised in the poems are those that seem important to the male speaker. In a sense, the women in the poems are merely objects defined by men. None of them speaks, so we have no idea of how they see themselves. We do not know whether they mind being compared to roses, stars or moons, or whether they place the same value upon the physical and emotional aspects that their suitors have chosen to highlight. Why should hair, eyes, teeth, breasts and legs define desirability? Some feminist critics would argue that contemporary society functions in a similar fashion. Notions of beauty and femininity, of love and of desire, which we take for granted, are often deeply conditioned by the expectations of a patriarchal society. Furthermore, women often accept these criteria as goals to which they should aspire. Indeed, attractiveness is often measured by the extent to which women are able to meet the idealized images promoted by advertising and the media.

WHAT ARE INTERPRETATIONS OF POETRY MADE OF?

Essays or paragraphs interpreting poems should contain at least the following elements (which will appear if you complete the paragraph above):

1 A statement of what you believe is the main idea or topic of the poem.
2 A discussion (often called analysis) of the language (including, but not only, figurative language) used by the poet to get the main idea across to the audience.
3 An evaluation of the poem, saying whether you think it 'works' or not and why.

TIME TO REFLECT

✪ In your view, what physical and emotional qualities make up the ideal man (or woman)?

✪ Do you think our sense of what constitutes the ideal man (woman) is determined by women (men)?

✪ Write a poem in which you attempt to flatter (or seduce) the ideal man (or woman).

. . . .

FOUR

CONTEMPORARY
POETRY: LOVE
AS DESIRE

LET'S LOOK AT a poem describing a woman's feelings for a man.

Christine Marshall is a South African poet. She was born and grew up in the Western Cape, but now lives in Gauteng. She writes 'bilingual poetry' in English and Afrikaans, trying to make the poems exactly equal in both languages.

IMAGINATION
Christine Marshall

When I was still bluffing myself
that I could love you
in my head only,
it amused me to think,
for all I knew
you could be a large magic doll
with a real face and hands
but a calico body
underneath your clothes.

When I had to acknowledge
to myself
you would never love me,
I smiled at you still
though desolation was shrinking me
at the thought
that I would never know
the shape of your feet,
the smell in the hollow of your neck,
and the feel of the skin
stretching down across your spine.

VERBEELDING
Christine Marshall

Toe ek myself nog wysgemaak het
dat ek jou slegs kon liefhê
in my kop,
het dit my amuseer om te dink,
vir al wat ek weet
kon jy 'n groot towerpop wees
met 'n regte gesig en hande
maar 'n katoenlyf
onder jou klere.

Toe ek aan myself
moes erken
jy sou my nooit liefhê nie,
het ek vir jou bly glimlag
al het verlatenheid my laat krimp
by die gedagte
dat ek nooit sou weet
wat is die vorm van jou voete nie,
die reuk in die holte van jou nek,
en hoe die vel voel
wat afwaarts oor jou ruggraat span.

Complete the following list of images by listing the words and phrases in the poem that refer to the body:

a large magic doll;	
a real face and hands;	

What is the poet saying about the object of her affections? We are going to read the images in the first stanza closely. They are

✪ a large magic doll
✪ a real face and hands
✪ a calico body
✪ underneath your clothes

Where context is concerned, the Afrikaans section of 'Imagination/Verbeelding' demands our attention. The poet has deliberately and carefully written her poem in two languages. The effect of this is to force the reader (if he or she can read Afrikaans as well as English) to think about the possibilities of poetry in two languages and to make connections between the two 'versions' of the poem. This practice flies in the face of the common idea that a version of a poem in one language cannot be equivalent to a version of the same poem in another (which I mentioned in relation to 'Fantasies of a Love-thief').

Now answer the following questions as a way of exploring possible meanings of the images in the first stanza.

✪ If a speaker refers to someone else as 'a large magic doll', what is implied about that person?

CLOSE READING

Interpreting poetry is conventionally based on a skill called 'close reading'. Close reading was popularized by the school of New Criticism, whose main proponents were I.A. Richards, René Wellek and Cleanth Brooks. It is a technique where you look very carefully at individual words and images, thoroughly exploring all their connotations and formal aspects such as metaphor, simile and irony. The main criticism of this kind of interpretation is that it does not pay enough attention to historical context.

✪ If the person is not real, is this unreality physical or emotional? Explain your answer.

✪ And what about his 'calico body' (calico is a kind of plain cotton material)? Does this imply anything about his sexuality?

Some connotations of these images are: dolls can be manipulated and controlled, placed wherever their owners want them to be. The speaker seems to suggest that she would like the man she is writing about to be life-sized ('large'), under her control (a 'doll'), but still able to come to life ('magic'). In an interview, Christine Marshall herself said that the reference to a 'calico body' makes her think of a child's doll with an elaborately life-like face and very appealing clothing; but once the clothes are removed, there are only plain white, shapeless tubes of cotton stuffed with cotton wool. This adds another dimension to our interpretation because it implies that the man appears to be real and human (as shown by his 'real face and hands') but he lacks some important qualities when one looks 'beneath' his clothing.

Looking at the stanza as a whole, you will notice that the description of the man as a doll is framed as an imaginary scene. The speaker says that she used to amuse herself by thinking of him this way, when she believed that she could love him 'in my head only'. Does this mean that:

✪ she loved his mind only and not his body?
✪ she did not speak about her feelings?
✪ her love for him did not include any desire?
✪ she wanted not to feel desire for him?

I would say that only the first possibility is really unlikely, because the phrase 'in my head only' suggests that the speaker is referring to her mind, not to his nature. The image of the man as a doll is therefore a fantasy that she has made up. It helps to convince her that he could be an unreal person, like a calico doll, with a body that does not have any sensations. (Could this be true? Or is it just a reflection of the way he treats her?)

TIME TO WRITE

Here we want you to use the technique of *close reading*, which we have just applied to the first stanza of 'Imagination/Verbeelding'. Use this approach to explore the images in the second stanza. They are strikingly different from those in the first. They have directly physical, indeed erotic, denotations and connotations. Explore these by following the guidelines below.

- ✪ List the parts of the body that are mentioned.
- ✪ What are the connotations of these parts of the body?
  What is the tone of the descriptions used by the speaker?
- ✪ Why is this tone used?
- ✪ How do these images relate to the image of the man as a large magic doll in the first stanza?
- ✪ What does 'desolation was shrinking me' mean?

Now write a paragraph explaining how the first and second stanzas are connected. You might like to investigate the use of connecting words, such as 'when' (both stanzas begin with 'when') and 'though'. What picture does the poem create of the relationship between the speaker and the man she is addressing? What is the role of the body in this relationship?

'Imagination/Verbeelding' shows us, among other things, that it is not only men who turn women into objects of desire: women speakers can also do the same to men (although this challenges and subverts conventional gender roles). It insists that the woman speaker is a desiring person, not only the object of desire. One of the most famous women poets to write about the experience of having a body and enjoying sexuality is Grace Nichols, whose poetry we are going to look at next. Her poem, 'Like a Flame', describes the sexual 'chemistry' between a woman speaker and a man.

● HOW DOES 'Imagination/Verbeelding' relate to the tradition of courtly love poetry and flattery that we looked at in the preceding section?

LIKE A FLAME
Grace Nichols

Raising up
from my weeding
of ripening cane

my eyes
make four
with this man

there ain't
no reason
to laugh

but
I laughing
in confusion

his hands
soft his words
quick his lips
curling as in
prayer

I nod

I like this man

Tonight
I go to meet him
like a flame

THEME

The word 'theme' is often used to refer to the subject or
topic of a literary work: what the work is about. It is also
occasionally used to refer to a work's 'universal' or moral
concerns.

This is a very simple-looking poem. It
doesn't 'look' like a love poem. It is not as
ornate and image-filled as the other
poetry that we have looked at in this
chapter. But, as we will find, its *theme*,
like that of many we have studied, is
desire. Let's begin by making a list of the
unusual phrases and images in the poem.
Complete this table:

Image from 'Like a Flame'	Connotations
Raising up	
my eyes / make four	
like a flame	

We are not going to investigate every line of the poem (as a critic who was reading the poem within a New Critical paradigm, and wanted to arrive at a comprehensive interpretation, might do). Instead, we are going to focus on the fifth stanza. It reads:

> his hands
> soft his words
> quick his lips
> curling as in
> prayer

If we fill in the grammatical gaps in this stanza, we get a much more conventional-looking claim: 'His hands are soft, his words are quick, his lips are curling as in prayer'. These are descriptions of a person who is gentle (he has soft hands), speaks easily, and whose mouth seems to be suited for devotion or worship (prayer). Here the speaker is telling us what attracts her to this man. She feels that he is a sensitive and articulate person.

This description creates a word-picture of the man that is unusual. Unlike some of the poems we have looked at in this chapter, the speaker of 'Like a Flame' does not only focus on the man's physical or sensuous attributes. The line 'his words [are] quick' tells us more about his behaviour than about his body. This is a strange attribute for someone to find attractive. Why

If you enjoyed 'Like a Flame', you might also like to read this extract from another Grace Nichols poem, 'Invitation'. This poem also deals with sexuality. Here it is:

FROM: 'Invitation'
Grace Nichols

> Come up and see me sometime
> Come up and see me sometime
> My breasts are huge and exciting
> amnions of watermelon
> your hands can't cup
> my thighs are twin seals
> fat slick pups
> there's a purple cherry
> below the blues
> of my black seabelly
> there's a mole that gets a ride
> each time I shift the heritage
> of my behind
> Come up and see me sometime

does the speaker like the fact that the man's 'words [are] quick'? When I was thinking about this, I wondered whether his ease with speech entertains her or makes her feel relaxed, and so encourages her to speak about herself.

Now let's look at the ending of 'Like a Flame'. It reads:

> Tonight
> I go to meet him
> like a flame

This stanza evokes the anticipation that the speaker feels about her plans to meet the man later. The word 'tonight', alone on a line, enables the reader to savour the pleasure she feels about their fast-approaching encounter. The word 'I' that begins the next line emphasizes her own sense of power: she is going to meet him, she is not just passively waiting for him to come to her. And the final line, 'like a flame', conveys the passionate sexual desire that the speaker is looking forward to enjoying with the man.

TIME TO WRITE

Write a paragraph evaluating Grace Nichols's poem, 'Like a Flame'. You might use the following steps in constructing your evaluation:

✪ State your opinion of the worth of the poem (this evaluation might include an interpretation of the poem as a whole);
✪ Mention some of the poem's strategies, such as images,

typography (page layout), line length or sound effects;

✪ Describe how these strategies work in the poem;

✪ Summarize what you most like or do not like about the poem.

You have just written an evaluative piece. As you reflect on the experience of writing it, think about how much interpretation went into your evaluation. Notice how you had to explore what you thought the poem meant at the same time as you commented on whether you thought it is successful or not.

· ·

PERHAPS YOU ARE ready to try writing a love poem. Below, we have given some approaches to writing poetry. You might find these helpful. You need to bear in mind, though, that the creative process is very individual. Each poet jogs his or her creative abilities by using the strategies that 'work'. These strategies differ widely amongst writers. So, as you write, you need to use any approach that sparks your own creativity.

FIVE

YOUR OWN LOVE POEM

 Strategy A

STEP 1: IMAGES

Write several phrases in the following format:

My love is like ....

My lover's body is like ....

My feelings are ....

STEP 2: PERSUASION

Complete the following sentences:

I want the person I love to ....

I will try to get his/her attention by ....

I might promise to ....

STEP 3: MAKING A POEM

Now make your ideas into a poem. You may use some or all of the following poetic strategies (the poems we have studied in this chapter use many of these, although few use all of them):

○ dividing the poem into stanzas
○ rhyme
○ repetition (such as a poetic 'chorus' at the end of each stanza)
○ regular rhythm
○ imaginative images

 **Strategy B**

A MENTAL PICTURE

Sit in a quiet spot and think of pictures and images that you associate with love. It does not matter if these are unusual or unconventional – another word for unconventionality is originality! When you have come up with a picture that 'feels right', that is, one that effectively evokes the idea of love, try to write it down.

SOME HINTS

When you try to 'translate' pictures in your mind into words on a page, you will probably find that a lot of time is needed for you to find the right words – that is, those that effectively express what is in your imagination. So be prepared to spend time drafting and re-drafting your poem.

SOME STRATEGIES

You may like to try some or all of the following strategies and techniques in making your 'word-picture' into a poem.

○ Divide the sentences into lines (follow your instinct here, or try reading the sentences aloud to see where your voice would pause).

✪ Divide your poem into stanzas (each stanza usually expresses a different idea).

✪ Use sound effects such as rhyme, or use words that sound similar to their denotative meanings, such as 'crash'. (This is known as onomatopoeia.)

Strategy C

CRAZY LEXICON

Go to the dictionary and extract six words at random. Write them down on a piece of paper. Now make a love poem that includes all six words. This will exercise your verbal ingenuity!

. . . .

IN THIS SECTION we have collected some poems that may transgress what are commonly thought of as 'normal' attitudes to sexual and romantic behaviour. Here is *The Oxford Advanced Learner's Dictionary* definition of 'transgress':

SIX

TRANSGRESSION

> **transgress** /trænz'gres; *US* træns-/ *v* **1** (*fml*) to go beyond the limit of what is morally or legally acceptable: [Vn] *transgress the bounds of decency.* **2** ~ (**against sth**) [V, Vpr] (*arch*) to offend against a moral or religious principle; to SIN[1] *v.*

All cultures have (often unspoken) conventions concerning what is sexually acceptable. Ideas about love, beauty and desirability also vary from time to time, society to society and from person to person. What is acceptable in one society or community may be forbidden in another.

✪ In your opinion, what makes someone beautiful or desirable?
✪ Do you think that these ideas about this are shared by most people?

● Do you believe that some sexual behaviour is 'right' or 'wrong'?

● How do you respond to sexual behaviour that goes against your moral beliefs?

● How do you respond to representations of sex (in movies, books and poems), especially when the behaviour represented seems morally unacceptable to you?

The *Oxford Advanced Learner's Dictionary* defines 'taboo' like this:

taboo /təˈbuː/ *n* (*pl* **taboos**) ~ (**against/on sth**) **1** a cultural or religious custom that forbids people to do, touch, use or talk about a certain thing: *the taboo against incest* ○ *tribal taboos* ○ *Death is one of the great taboos in our culture.* **2** a general agreement not to discuss or to do sth: *There's a taboo on smoking in this office.*
▶ **taboo** *adj* forbidden by a taboo: *Sex is no longer the taboo subject it used to be.* ○ *Any mention of politics is taboo in his house.*
■ **taˈboo words** *n* [pl] words that are often considered offensive, shocking or rude, eg because they

The idea of conventionally accepted behaviour implies that certain things are taboo or forbidden. The poetry that we are going to read in this section deals with sexual practices that, in various ways, transgress the norms society usually accepts. While people commonly encounter sexually explicit language and behaviour in the mass media and in life, encountering these in literary texts prescribed for study is less usual. For this reason, you may find that the material touches on areas you find sensitive. Some of the poetry aims to shock the reader, so do not be alarmed if you find yourself responding with agitation. When this happens, stop and think about why you are shocked. Is it because the poetry violates what is 'right' or, at least, what you *think* is 'right'?

A key question is why writers choose to write about sexual behaviour that might offend readers. Are they trying to shock or to be sensational? Are they hoping that if they write about matters that are usually considered shocking, they will have a better chance of being considered original and therefore having their work published? Texts about sexual 'transgression' implicitly ask readers to think about why we take certain behaviours for granted and automatically reject others. This leads us to ponder the origins and construction of the norms that operate in our society and culture. Many such norms come from deeply held systems of belief, which consciously and unconsciously influence our choices about what is right or wrong. Others may be the result of stereotypes, which in turn give rise to oppression (for example, sexist ideologies). As long as we believe that there is only one type of 'normal' or 'good' love and sexuality, we risk reducing all others to deviance. By focusing on sex, the following poems challenge readers to reflect on norms they might have taken as 'given'.

As you read through each of the poems we have selected, you should ask yourself these questions:

- ❂ Why has the writer chosen to write so explicitly?
- ❂ Is the poem transgressive in any way? And, if so, how?
- ❂ How do you respond to the ideas and descriptions in the poem?
- ❂ In your opinion, does the poem exceed the limits of what is acceptable?

PORNOGRAPHY VS. EROTICISM

D.H. Lawrence's novel *Lady Chatterley's Lover* (which was first printed privately in Florence in 1928) raised a number of questions about eroticism and literature. Lawrence's poetic but detailed descriptions of sex and his relentless use of four-letter words caused the book to be unpublishable in full in England until 1960 when Penguin Books risked printing a complete text. They were prosecuted under the Obscene Publications Act of 1959 and acquitted after a legendary trial during which eminent authors (including E.M. Forster) appeared as witnesses for the defence. The book's victory had a profound effect on both writing and poetry publishing in later decades.

Other literary works have also been the subject of court trials because of their representation of sexuality. Famous examples include Radclyffe Hall's classic lesbian novel, *The Well of Loneliness*, and James Joyce's *Ulysses*.

All the poems in this section are *optional*. If you find them offensive, remember that you do not have to read them. It is not our intention to force you to examine the representation of practices, lifestyles or behaviours that make you feel uncomfortable.

In the first poem, the speaker argues that sexual intercourse is an invasion of the female body; that it is boring and should be done away with. She advises that the 'solitary act' (i.e., masturbation) is more convenient and satisfying.

AGAINST COUPLING
Fleur Adcock

I write in praise of the solitary act:
of not feeling a trespassing tongue
forced into one's mouth, one's breath
smothered, nipples crushed against the

the solitary act:
masturbation

ribcage, and that metallic tingling
in the chin set off by a certain odd nerve:

unpleasure. Just to avoid those eyes would help –
such eyes as a young girl draws life from,

vegetal: like a plant

listening to the vegetal
rustle within her, as his gaze

polypal: like the arms of an octopus

stirs polypal fronds in the obscure
sea-bed of her body, and her own eyes blur.

There is much to be said for abandoning
this no longer novel exercise –
for not 'participating in
a total experience' – when
one feels like the lady in Leeds who
had seen *The Sound of Music* eighty-six times;

or more, perhaps, like the school drama mistress
producing *A Midsummer Night's Dream*
for the seventh year running, with
yet another cast from 5B,
Pyramus and Thisbe are dead, but
the hole in the wall can still be troublesome.

I advise you, then, to embrace it without

encumbrance: hindrance or impediment

encumbrance. No need to set the scene,
dress up (or undress), make speeches.
Five minutes of solitude are
enough – in the bath, or to fill
the gap between the Sunday papers and lunch.

✪ Why does the speaker refer to sexual intercourse as 'coupling'?
✪ What reasons does the speaker offer for her rejection of intercourse?
✪ What do you think of the speaker's arguments?
✪ Do you find the poem offensive? Is masturbation a suitable or unsuitable topic for literature?

The tone of 'Against Coupling' is not especially serious. The

LOVE, SEX, BODY

speaker gently makes fun of the act of lovemaking (and the tradition of representing it), arguing that these works and the activities they represent can become invasive and boring. At the end of the poem, she turns her parodic gaze on herself, suggesting that, in fact, the alternative to lovemaking can, itself, be rather trivial and mundane. Look at the way the poem emphasizes this through the use of everyday language.

Perhaps what surprises us about 'Against Coupling' is that a female writer deals with a subject that is not usually spoken about (particularly by women in public) in a very matter-of-fact or ordinary way. The following two poems also challenge our ideas of what is acceptable. The first poem, '"Vagina" Sonnet', raises interesting questions about gender and the language of poetry, while the second poem rejects the objectification of feminine beauty.

'VAGINA' SONNET
Joan Larkin

Is 'vagina' suitable for use
in a sonnet? I don't suppose so.
A famous poet told me, 'Vagina's ugly.'
Meaning, of course, the *sound* of it. In poems.
Meanwhile he inserts his penis frequently
into his verse, calling it seriously, 'My
Penis'. It *is* short, I know, and dignified.
I mean of course the sound of it. In poems.
This whole thing is unfortunate, but petty,
like my hangup concerning English Dept. memos
headed 'Mr/Mrs/Miss' – only a fishbone
In the threat of the revolution –
a waste of brains – to be concerned about
this minor issue of my cunt's good name.

A NUDE BY EDWARD HOPPER

for Margaret Caul

Lisel Mueller

The light
drains me of what I might be,
a man's dream
of heat and softness;
or a painter's
– breasts cosy pigeons,
arms gently curved
by a temperate noon.

I am
blue veins, a scar,
a patch of lavender cells,
used thighs and shoulders;
my calves
are as scant as my cheeks,
my hips won't plump
small, shimmering pillows:

but this body
is home, my childhood
is buried here, my sleep
rises and sets inside,
desire
crested and wore itself thin
between these bones –
I live here.

Douglas Livingstone's poem 'Giovanni Jacopo Meditates (on an early European navigator)', like Fleur Adcock's 'Against Coupling', is about a sexual act. It describes the lover's body in ways that are explicit and sensual. However, it does this by using the metaphor of a journey, undertaken by 'Sir Tongue,' who is the person or object addressed by the poem's speaker. In using the language of travel and discovery, Livingstone – a twentieth-century South African poet – mimics a style of poetry that belongs to the Elizabethan tradition when the voyages of discovery to undiscovered territories had an enormous influence on literature, including love poetry. As is the case with much love poetry from this earlier era, Livingstone represents the body of the beloved in terms of geography, navigation and mapping.

GIOVANNI JACOPO MEDITATES
(ON AN EARLY EUROPEAN NAVIGATOR)
Douglas Livingstone

I adjure thee, Sir Tongue: Be Firm. Be Indiscrete.
Cast off. Your Journey starts from her slightest Toes.
Set Sail upon the Creases of her Feet,

Up, over her slim Ankles; perhaps at Sea
On choosing which Course or Calf: follow your Nose;
Arrive to linger on the Pool behind each Knee.

Here, you may gather Wits & Breath to sound
Your Strength for that Expanse that lies ahead:
Those Seas reach on, each with its Round & Mound.

Tack up the Backs of her slippery Thighs
– Lash yourself to the Helm: lose not your Head,
But keep it down. Round, in between her Nates rise

To pilot the Archipelago of her Spine.
Fare to her Shoulders from her Arms' lax Sweep;
You'll reach her Nape: this be your Journey's Shrine.

But if, Sir Tongue, your Exploration's seen
A Failure by her mere Murmuring in her Sleep,
As fitting Worth and worthy Fitter to our Queen:

Why, you must coax her over. Perhaps a Tease
Is here to stretch your Voyagings. So buckle to;
Return; & make Eyelids & her Mouth unfreeze.

Down over Chin & Throat to Armpits you'll be sent,
& up those Sun-Tipped Capes from whence a Country-View

nates: buttocks

archipelago: a chain of islands

nape: back of the neck

Spreads below. Coast down to her soft Belly's Dent.

Here, you may pause to ease your Rig and Sails.
Cruise in widening Circles until intervenes
That Continent's sweet Harbour from the South-West Gales.

Drop Anchor in this most redolent of Coves,
& taste for yourself Nectarines, Tangerines,
Pineapples, Grapes, Avocados, Paw-Paws, Cloves.

Now you may rest. With modest Stirrings sit
Slaking & sluicing all your long Journey's Care.
Survey the Port, Sir Tongue, where best you may refit.

Remain alert for the Storm that overtakes
Seismic Tremors with unbridled Waves. Beware:
There will be Tumults when her fevered Body wakes.

✪ Imagine that the speaker is talking to you, in other words
that you are 'Sir Tongue'. How would you feel about under-
taking the journey described in the poem?

✪ Now place yourself in the position of the woman whose
body is being explored. How would you feel if your body was
spoken about in this way?

✪ Since it shows the woman's body as an object for pleasure, is
the poem sexist?

✪ In what way is this poem transgressive?

Under the guise of sixteenth-century courtly love poetry (using
flattery to exaggerate the beloved's charms), 'Giovanni Jacopo
Meditates (on an early European navigator)' is remarkably
explicit about the physical features that are caressed by 'Sir
Tongue' on 'his' journey up and down the woman's body. Many
of the words and phrases used by the speaker have associations
with exotic places: for example, 'Archipelago', 'Sun-Tipped Capes'
and 'that most redolent of Coves' (referring to the woman's vagina). The references to the tropical fruits that grow in the 'redolent Cove' give the impression of lush growth and delicious tastes
and textures.

- Do you find references to anatomy in love poetry offensive? If so, do these become less offensive when they are disguised, as in 'Giovanni Jacopo meditates (on an early European navigator)'? Explain.
- Should the speaker in the poem be more concerned with the woman's inner qualities and feelings than with the contours of her body?
- If he were, would you feel that his affections are more sincere?
- Are we more convinced by poetry and discourse that is directed towards the mind and heart of the beloved than towards his or her body? Explain.

The following extract from a prose-poem by Alta is also very explicit about body parts and erogenous zones. Is it sincere? Does it represent acceptable literary expression?

I WOULD WRITE

Alta

i would write of such a love, such a body, such a person
that loves me so. i would write, & be glad, & forget that i sit
alone in this room, with no one to come see me.

     remember instead the ringing telephone, the soft low
voice, the promise of naked warmth. i want my breasts
against his chest. i want his hands on my hips, moving
smoothly down the curve of my hips as i breathe in his
smell. i am tired of people finding each other ugly. i want to
sing the beauty of all bodies, of the joy of touch, the warmth
& softness of heavy bodies, the tight energy of hard bodies,
the soft, melting breasts of mothers; the small buds of
breasts of girls. the soft curled cocks of men before they
want me; the way they fill & rise to fill me up with their
love, the curve of the butt below the balls as he lies on his
back, smiling at me, his cock moist & hot. his smooth
brown chest, belly flat below his risen cock. the beauty of
my soft white breasts coming down on him, my full thighs
around his tiny hips. the textures, the colours, the hard
places & the soft places, the strength of our bodies as we
come together ...

The next poem also offers a direct and challenging representation of lovemaking. It is a fairly conventional poem, depicting two people who are sexually involved. However, the expression of rising passion and the accompanying feelings of love are unmistakable.

STRONG THIGHS ASTRIDE MY CHEST
Caroline Griffin

Strong thighs astride my chest
your body presses wet against me
draws up passion as we meet and
yes   your hands press down my palms

I watch you seek me out
this powerful mother who
licks her child with passionate tongue
whose urgent fingers
        touch my lips
        open my mouth
        hold your breast to me –
this is the freedom of wild mothering
to choose
to reach for a hand we don't let fall.

And we can hold together in a circle
our bodies cradle an energy
which spills in sweat in breath in cries
and choose to see our faces as we flow.

- What was your first reaction to this poem?
- How does the use of pronouns differ in the first stanza and the last one? What is the effect of this change?
- What is the effect of using punctuation in the final stanza?
- Do you find the intimate descriptions of the lovers' bodies and action unusual, appropriate, shocking or attractive?
- Are there images in the poem that seem unconventional?
- Would your response change if you were told that the poem describes lesbian sexual activity? Why?

For a woman to identify herself as a lesbian, whether openly or in private ('in the closet'), is a highly transgressive act as it implies that she does not wish to be part of the dominant order of sexuality, where women are 'supposed' to be fulfilled by men. Therefore lesbians exist, by virtue of their sexual choices, 'outside' of 'mainstream' sexuality. The poem emphasizes this through the repetition of the word 'choose'.

Until recently, gay men have also existed outside normal social life. Today, though, in some contexts, attitudes towards homosexual behaviour have become more tolerant. Moreover, gay people have actively campaigned against their discrimination. 'Gay Epiphany' unashamedly celebrates the male sexual anatomy. It is a list of anatomical terms referring to the penis and reproductive body parts of men. The repeated use of 'o' is an almost ritualistic invocation of praise and celebration of these parts of the body. The speaker focuses relentlessly on the sexual organs. In this way, the poem implies that male homosexual desire can be as physically involved and detailed as heterosexual desire. This is a strongly political claim since men's sexual desire for one another has generally been marginalized, made invisible, or even seen as deviant and illegal. By placing his admiration for his lover's sexual organs in the foreground of the poem, the speaker makes an argument giving male homosexual desire the same status as heterosexual eroticism.

GAY EPIPHANY
James Mitchell

epiphany: a manifestation
of the divine

> *In a culture where the aesthetic experience is denied*
> *and atrophied, genuine religious ecstasy rare, intel-*
> *lectual pleasure scorned – it is only natural that sex*
> *should become the only personal epiphany of most*
> *people ...*
>
> (Gary Snyder)

o sperm, testes, paradidymus! o scrotum, septum, and
    rectum!
o penis! o prepuce, urethra!

o prostate gland! o Dartos muscle! o spermatid and
    spermatocyte!
inguinal canals! seminal vesicles! seminiferous
    tubules!
prostatic uricles! efferent ductules! testicular lobules!
o male germinal epithelium!

o symphysis pubis! tunica albuginea! vasa efferentia!

corpus cavernosum et spongiosum! o ampulla of vas!

o meatus and bulb! o cutaneous dorsal vein! o lobulous
membranous convoluted pouch!

o glans penis, slightly bulging structure at the distal
    end of the penis!

o Cowper's glands, secreting a slimy substance which
    functions
as a lubricant!

• DO you think
that romance and
desire between
people of the
same sex is worthy
of celebration in
literature in the
same way as
heterosexual love?

o seminal fluids, functioning to suspend and protect the
delicate spermatozoa during their stay within the male
    body!

o epididymis, a tube-like structure commonly attain-
    ing a length of up
to twenty feet!

Boy, at the lovely tip of your external urethral orifice,
all my poetries
terminate

The impact of 'Gay Epiphany' lies in the final lines:

Boy, at the lovely tip of your external urethral orifice,
all my poetries
terminate

The speaker implies that he has nothing more to say, that words
and even poetry run out when he is confronted with the tip of
his lover's penis. This is a clever poetic strategy. Having demonstrated the speaker's capacity for effusive and detailed praise of
his lover's anatomy, it implies that his admiration for this particular body part is so great that he cannot articulate it in words.
The line is also mimetic of the act of climaxing. The poem ter-
minates on the word 'terminate'. Apart from praising the lover's
anatomy, the poem also satirizes the excessive attention granted
to the physical body (by society in general, including gay men).

The extensive use of anatomical terminology deliberately draws attention to the biological and physical features that have replaced the 'inner person' as the true objects of love and desire. Moreover, there is something slightly ridiculous about finding the ultimate experience of sexual perfection at the tip of someone's penis. Ironically, this celebration of homosexual desire may also be a critique of it.

In this section we have looked at a number of 'transgressive' poems dealing with love and sexuality. These poems challenge and undermine conventional notions of what is acceptable in romance and loving. Perhaps some of them made you feel uncomfortable and this may well have been the writer's intention. We have tried to include thought-provoking poems. We have not included any poems in the expectation that you should adopt forms of sexual morality that make you uncomfortable. Sex, sexuality and sexual behaviour are inescapable parts of contemporary life. Trends in the mass media, together with the impact of the HIV/AIDS pandemic, have ensured that topics that were traditionally discussed by adults in the privacy of their bedrooms have become matters of conversation among a younger audience. It is in this context of 'openness' that we have asked you to explore your responses to literary representations of love, sex and the body, without setting limits or directing your reactions.

The final section of this chapter does set limits and take a stand. Until now, the sexual behaviour we have dealt with involves consent or choice. People can decide for themselves whether they wish to take part in it or not. Rape is not consensual. It does not involve choice on the part of the recipient. It is an act of extreme violence that can only be condemned.

● WHAT limits in terms of subject matter and explicitness would you set if you were being asked to draw up a reading list dealing with a topic such as love, sex and the body? Explain why.

SEVEN

RAPE

RAPE IS DEFINED by *The Oxford Advanced Learner's Dictionary* like this:

> **rape¹** /reɪp/ *v* to force sb to have sex when they do not want to: [Vn] *She claimed she had been raped.*
> ▶ **rape** *n* **1(a)** [U] the crime of raping sb: *to commit rape* ○ *Rape is on the increase.* ○ *rape victims.* (**b**) [C] an instance of this: *a rise in the number of rapes.* **2** [U] (*fml*) the act of destroying or spoiling sth: *the rape of our countryside* (eg by building on it).
> **rapist** /ˈreɪpɪst/ *n* a person who commits rape: *a convicted rapist.*

THE current legal definition of rape in South Africa is: 'Unlawful, intentional sexual intercourse with a woman without her consent.' The definition is being reviewed at the moment and new definitions are being proposed.

✪ What are the limitations of this definition?

Rape is an act of violence, rather than one of desire. It is a very traumatic experience for the victim, especially since rapists often try to justify it with allegations such as 'she provoked me'; 'she led me on'; or 'she dressed in such a sexy way that I thought she wanted sex'. Also, although it is a crime that carries a heavy penalty, it is often extremely difficult to prove in a court of law. Because of this and because legal trials recall the traumatic experience, many women who have been raped feel reluctant to prosecute the offenders. Thus, rape becomes a secret, often shameful trauma with damaging and long-term psychological effects.

Rape is not commonly dealt with in literature, especially not in poetry that relies heavily on metaphor and imagery. It is difficult to use metaphorical language when writing about an act of sexual violence.

HELP FOR RAPE VICTIMS

If you have been a victim of violence or rape, you may want to consider counselling or another form of help. Look up the telephone numbers for the following organizations in your local telephone directory:

Lifeline
Rape Crisis
People Against Women Abuse (POWA)

So how have poets reacted to rape? How do we react to literary representations of rape? One response amongst many possible ones to the prevalence of rape is anger, expressed in distinctly 'unpoetic' language, as in Jayne Cortez's poem, 'Rape', which we are going to read next. Bear in mind, though, that 'Rape' gives only one perspective on this crime. Remember that you may choose not to read this poem (or this section) if you find its subject distressing.

RAPE
Inez Garcia, Joanne Litte – Two Victims in the 1970s
Jayne Cortez

What was Inez supposed to do for
the man who declared war on her body
the man who carved a combat zone between her breasts
Was she supposed to lick crabs from his hairy ass
kiss every pimple on his butt
blow hot breath on his big toe
draw back the corners of her vagina and
hee haw like a California burro

This being war time for Inez
she stood facing the knife
the insults and
her own smell drying on the penis of
the man who raped her

She stood with a rifle in her hand
doing what a defense department will do in times of war
And when the man started grunting and panting and
wobbling forward like
a giant hog
She pumped lead into his three hundred pounds of
shaking flesh
Sent it flying to the Virgin of Guadalupe
then celebrated day of the dead rapist punk
and just what the fuck else was she supposed to do?

JAYNE CORTEZ is a contemporary performance poet. The impact of her poetry comes from its being read aloud.

crabs: sexually transmitted parasitic lice

burro: a donkey used for carrying heavy burdens

And what was Joanne supposed to do for
the man who declared war on her life

Was she supposed to tongue his encrusted
toilet stool lips
suck the numbers off his tin badge
choke on his clap trap balls
squeeze on his nub of rotten maggots and
sing god bless america thank you for fucking my life away

this being war time for Joanne
she did what a defense department will do in times of war
and when the piss drinking   shit sniffing guard said
I'm gonna make you wish you were dead black bitch
come here
Joanne came down with an ice pick in
the swat freak motherfucker's chest
yes in the fat neck of the racist policeman
Joanne did the dance of the ice picks and once again
from coast to coast
house to house
we celebrated day of the dead rapist punk
and just what the fuck else were we supposed to do

This is an enraged poem, written in response to the phenomenon of rape. You may be shocked by the fury the poem expresses. But if we look at the following aspects, we find that it has a
carefully constructed argument concerning rape and women's
responses to it.

The speaker represents the bodies of the two male rapists as
ugly, taking care to make sure the reader does not sympathize
with them. (Which words and phrases show the ugliness of the
rapists?) The speaker also uses military images, which demonstrate how rapists misuse their bodies by turning them into
weapons of war. Make a list of the words and phrases used to create images of the rapists' bodies as repulsive.

The rape victims – Inez and Joanne – take revenge on the
rapists in a very violent way. The poem presents their retributive

violence as justified; but their actions raise the question of whether violence in revenge is justifiable or not. What do you think? Should rapists be treated in the way the poem represents?

Some phrases – 'day of the dead rapist punk' and 'just what the fuck else was she / were we supposed to do' – are repeated. These are very powerful 'What ... else was she supposed to do' is a rhetorical question that implies that violence is the only appropriate response to rape. Does the poem convince you of this? Or should victims try to deal with their trauma through legal action, without trying to get revenge through violent means?

There are a number of images relating to war, weapons and violence. These make the reader realize that rape is not a sexual or romantic phenomenon, but has to do with brutality and the misuse of power.

This is a performance poem: it is meant to be 'performed', read or recited aloud to an audience. This makes the poem even more of a call to action than if it were meant only to be read as words on a page because the voice of the speaker can emphasize the angry tone of the poem. Further, the repetition of certain phrases and expressions in the poem gains impact from being read aloud. (You may wish to try performing the poem yourself to explore this aspect.)

As we said before reading the poem, Jayne Cortez offers one perspective on rape. Some others are sympathy for victims, and sadness that such inhumanity and cruelty exist in human relationships. What do you think is the appropriate response to rape?

• HOW does the form of the poem relate to its content? Do you think it would be possible to express the same sort of fury in a sonnet.

RAPE POEM
Marge Piercy

There is no difference between being raped
and being pushed down a flight of cement steps
except that the wounds also bleed inside.

There is no difference between being raped
and being run over by a truck
except that afterward men ask if you enjoyed it.

There is no difference between being raped
and being bit on the ankle by a rattlesnake
except that people ask if your skirt was short
and why you were out alone anyhow.

There is no difference between being raped
and going head first through a windshield
except that afterward you are afraid
not of cars
but half the human race.

The rapist is your boyfriend's brother.
He sits beside you in the movies eating popcorn.
Rape fattens on the fantasies of the normal male
like a maggot in garbage.

Fear of rape is a cold wind blowing
all of the time on a woman's hunched back.
Never to stroll alone on a sand road through pine woods,
never to climb a trail across a bald
without that aluminium in the mouth
when I see a man climbing toward me.

Never to open the door to a knock
without that razor just grazing the throat.
The fear of the dark sides of hedges,
the back seat of the car, the empty house
rattling keys like a snake's warning.

The fear of the smiling man
in whose pocket is a knife.
The fear of the serious man
in whose fist is locked hatred.

All it takes to cast a rapist to be able to see your body
as jackhammer, as blowtorch, as adding-machine-gun.
All it takes is hating that body
your own, yourself, your muscle that softens to flab.

All it takes is to push what you hate,
what you fear on to the soft alien flesh.

To bucket out invincible as a tank
armoured with treads without senses
to possess and punish in one act,
to rip up pleasure, to murder those who dare
live in the leafy flesh open to love.

Political poetry such as 'Rape' and 'Rape Poem' demands a response from the reader: you cannot sit on the fence, but have to make choices about your views. Furthermore, some views entail political actions, such as joining an anti-rape activist group. Whatever your response, you need to think seriously about social problems such as violent crime and the misuse of power.

In the next part of this book, we are going to explore issues of power – its use and abuse as well as ideas on its revolution – in more detail.

PART THREE

Power

OVERVIEW

The final two chapters of *Love, power and meaning* explore the idea of power. Although we all have an intuitive grasp of what power means, it is not an easy concept to pin down. Not only does the word have many meanings, but the ways in which power affects our lives are also often complex and subtle. We may not recognize that most of the structures, ideas and actions that we take for granted as normal are linked to power. Power is everywhere: from the intimate arrangements of our family lives to the ways that powerful nations influence one another; from personal choices about what is right or wrong and individual issues of what to see, read or hear to decisions on holding elections, punishing criminals or waging destructive war. It is both the pervasiveness of power and the manner in which it is often hidden that makes it so fascinating.

Chapter five, 'Politics, language and society', invites you to investigate the social manifestations of power. Here we are most interested in the relationship between power and what people call 'politics'. Initially, the issues we'll deal with fall easily into a political category. For instance, we ask: Why do people allow themselves to be governed by authority? Who should govern? What form should government take? Later on, though, Chapter five asks us to extend our understanding of politics by examining the roles that language and knowledge play in society.

In Chapter six I want to look at the politics of reading, the politics of writing and the politics of writing about literary works (what is known as 'literary criticism'). Although the discussion deals with drama, it is applicable to criticism as a whole. Our aim is to give you an example of the actual activities involved in making sense of a text, while considering the assumptions and implications that underlie these activities. Furthermore, by opening up the critical process we hope to shed new light on the question that prompted this book: What is the value of literary and cultural study in contemporary society?

5 Politics, language and society

NO PERSON LIVES out his or her entire life in complete isolation from other people. To seek companionship, to associate with others, is a fundamental aspect of what it means to be human. Human beings are social creatures, and as soon as two or more people interact with each other we have the skeleton of a communal structure that we call society.

This communal social structure may be small – a marriage partnership, for instance. Or it may be large, comprising all the individuals in a neighbourhood, a city or a nation. But whatever its size, any social structure is likely to have certain characteristics. First, the individuals in a relationship will have particular roles that are mutually understood and defined. Secondly, these social roles probably will be based on codes of behaviour that are accepted as natural and necessary for the smooth functioning of the social arrangement. Finally, the social grouping will try to protect its particular interests and to prosper. In a family, for instance, parents and children have very distinct roles, with clearly defined responsibilities and duties. Traditionally, parents are expected to love and nurture their children, while children are expected to respect and obey their parents. When the participants overstep the boundaries of their roles, this behaviour is condemned. Neglectful parents may not be allowed to live with their children, while badly behaved children are likely to be punished. At the same time, by forming a defined unit, the members

ONE

PERSONAL
POLITICS

of a family are able to support each other in times of difficulty and to share accumulated resources in times of plenty. Yet, even such a close-knit community as a family is influenced by power.

TIME TO REFLECT

Think about your own family and answer the following questions:

- How is your family made up? Does it consist only of parents and children or does it include other relatives?
- Do all members of your family have an equal say in the decisions of daily life?
- Do all members of your family contribute equally to the successful maintenance of the group as a whole?
- Are individual members of your family expected to carry out particular tasks? If so, who does each task?
- Who in your family is likely to take important decisions and make rules?
- Who carries out the punishment when rules are broken?
- To what extent are the children in your family allowed to express their opinions and preferences?
- How does your family resolve any conflicts that may arise between its members?

I suspect that your answers to these questions reveal the distribution of power in your family very accurately. I would also guess that the ways in which your family members interact with each other are quite complex. Factors such as the composition of the family, its source of wealth, the age of the various family members, religious belief and educational background, among others, all influence the amount and type of power that is given to individuals.

- What does your community or society regard as the ideal family?

✪ From your observations, what constitutes a typical family? Are there differences between the typical family and the social ideal?

✪ Now ask an elderly person to describe family life and family structures when he or she was young. Write down points of difference and similarity between now and then.

THE HISTORY OF THE FAMILY

Until about fifty years ago, the typical family in a Western context was made up of a mother, father and any children that they might have. Grandparents, uncles and aunts, while regarded as relations, were not considered to be part of the unit comprising an individual household. Across Europe, Britain and North America the close family unit, sometimes referred to as the nuclear family, with the father as the head and the breadwinner, was the norm. By contrast, in many other contexts, extended family arrangements dominated. Here, divisions of authority were complicated by factors including tradition and age. In other contexts, it was not unusual for a polygamous husband to have several wives, each with her own rank. Less common, though not unknown, were polyandrous societies where wives wielded the greatest influence and were entitled to more than one husband.

Recently, in many contexts, traditional notions of the family have fallen away. More and more families have a single parent and in some instances same-sex couples are legally entit-led to raise children. Elsewhere, the extended family has buckled under the pressure of capitalist economies or the individual's pursuit of greater autonomy. Similarly, polygamy has been challenged and polyandry is on the wane. Given the wide variations in family structure and their susceptibility to change, it is difficult to define what makes a typical family. It is even more difficult to argue for an ideal family structure or to claim that certain divisions of power and responsibility among family members are preferable to others.

Of course, the family is not the only social group to which people belong. For most of our lives, we are part of far larger social organizations, each existing for a defined purpose and each operating according to its own set of principles and rules.

✪ Do you belong to any of the following social groups?

a sports team	a political party or a workers' union	a band or choir
a school, college or university	an administrative department	a social club
a company carrying out business or providing a service	a religious organization, such as a church or mosque	an association or committee

✪ What is your role in the social organizations to which you belong?
✪ Who leads the group and how is he or she chosen?
✪ What rules govern the members of the social group/s?
✪ Who decides on the rules?
✪ What action is taken against people who break the rules?
✪ Who enforces this action?

As we all know, if we belong to a group we need to set aside aspects of our individuality. We have to be prepared to coexist and co-operate with other people. In small groups such coexistence might depend merely on shared understandings and interests. However, in larger social organizations matters of coexistence are not left to chance. Rather, they are actively regulated by legal mechanisms. Thus, employees may have to sign a code of conduct and employers will enter into contracts that set out the conditions of service for employees. Similarly, before being allowed to join a political party or a trade union, prospective members may have to meet particular criteria or even pledge their allegiance to the ideals of the organization. In these instances, the social association is a voluntary one. People can choose whether or not to belong.

POWER

In many other important instances, freedom of choice is limited. As inhabitants of a town or city and as citizens of a country, individuals may find themselves subject to conditions and policies that are not of their choosing or making. In addition, at this level of society, the roles and rules of citizenship are usually encoded by a complex legal system, promoted by social and educational institutions and enforced by active policing and an elaborate system of punishment for offenders. This is usually what comes to mind when we think of power and of politics.

Individuality and Government

Measured against the power of a nation's government, the power of an individual seems insignificant. Yet, in most modern societies, individuals are able to exercise their power in the political arena. Obviously, in democracies, people can vote for the party or candidate of their choice. They may even actively campaign for a particular candidate or policy. Moreover, in most societies, ordinary citizens may seek political power for themselves by running for office at various levels. If these opportunities are denied, then people (in South Africa and in many other countries) have shown opposition to unjust rule through acts of civil disobedience and armed revolution.

Yet, it would be foolish to think that power and politics operate only at the highest levels of social organization. Any careful examination of human interaction, even if it is confined to the intimate details of a relationship between two people, will show that it is riddled with issues of power and politics. Exerting your will or influence, expressing a preference, expecting particular types of treatment and behaviour, all of these involve power and any exercise of power has political implications. To think of politics as referring only to what happens in parliament limits the meaning of the word too severely. Most modern thinkers understand politics as a broad term covering the ways in which people organize themselves in society. Men who expect their wives to be subservient to them are acting in a way that is no less political than superpowers who threaten to bomb their opponents into oblivion. If power is everywhere, so is politics.

Whether we are active participants in political processes or whether we passively observe the social interactions of our societies, we cannot escape politics and power. As social creatures, we are unable to remove ourselves from the results of living in society. Not to act, not to hold a view or take a stand on an issue is as much a political stance as to act or to adhere to the programme of a political party. Silence can be as damning as speech, if not more so.

Of course, the pervasiveness of power and politics should not lull us into believing that they are constant entities. The mere fact that societies and social values vary greatly across time, space and culture shows that they are unlikely to be consistent or stable.

Diversity and variation suggest that it may be difficult, if not impossible, to reach universal agreement on the ethics of power. Questions of legitimacy and illegitimacy, of morality or immorality, what is right or wrong are always relative to context, circumstance or intention. It could be argued that, in certain circumstances, society or the State has the right to pursue whatever line of action serves the interests of the majority. Rulers or governments may lie, cheat, torture and murder in order to preserve order, to ward off a threat or to protect innocent civilians. This is the pragmatic view of power. On the other hand, it can be argued that societies should always aspire towards the ideal exercise of power. In this view, there should be clear distinctions between legal and illegal authority, and firm boundaries that limit the powers of the State and the power of individuals.

TIME TO DISCUSS

✪ Do you think that the end justifies the means – that as long as your aims are honourable, the methods that you use to achieve them are unimportant?
✪ What set of principles characterizes your society?

POWER

✪ Do you think that it is possible to govern a society according to a set of ideals or is it necessary for laws to be practical?

✪ If you were given the power to create an ideal society, what would it be like?

. . . .

◯ Social Beginnings

For all recorded history, humans have lived in groups. In fact, palaeontologists (scientists who study extinct and fossil animals and plants) have shown that the ancestors of modern humans have lived communally for millions of years. The advantages of forming social structures must outweigh the disadvantages. Furthermore, even in the most distant past and in the smallest groupings, social living must have given rise to particular power relationships between the participants. Politics emerged in these first rudimentary or basic societies. In this section we are going to take an imaginary journey into that distant past. The purpose of our journey will be to think about our political origins.

Let's turn back the clock!

Time: 50 000 years before the present
The Place: A Neanderthal settlement somewhere in Europe
The Situation: The invasion of territory by another clan

WHO OR WHAT WAS NEANDERTHAL MAN?

Neanderthal man was a prehistoric being who lived in specific parts of Europe, Asia and Africa from about 100 000 to 35 000 years ago. Although the exact link between Neanderthal man and modern human beings (*Homo Sapiens*) is scientifically controversial, many scientists classify Neanderthals as an earlier sub-species of *Homo Sapiens.*

Fossil evidence shows that these people stood about 1,57 metres tall and walked fully erect. They had heavy bones, slightly curved limbs, big brow ridges and powerful teeth. Their brains were as large as those of modern humans.

The climate in which they lived was much colder than today. Nevertheless, stone tools and animal bones found with the human fossils show that the Neanderthals were successful hunters in their subarctic environment. They lived in simple temporary shelters, and sometimes also in caves, while following the animal herds. Significantly, archaeologists have found evidence that at least some Neanderthal people buried their dead with great care — the bodies were sometimes laid to rest along with stone tools, animal bones and even flowers.

Let us try, through the power of imagination, to enter the mind of a member of the clan whose territory is threatened by marauding enemies. Draw near as we overhear his outrage:

This is our land. This is our hill. These are our caves. We can see the countryside for miles around as our enemy approaches. Our lives are bound up with its rhythms and its seasons. We hunt in the spring and the summer when our deer come to drink at the river, we salt our meat in the autumn for the time when we huddle in caves, taking shelter against drifting snows. Fire warms our caves when icy mists blacken our days. It is unthinkable that we should live anywhere else, for the gods gave us this place for our children and our children's children. They guard it with their spirits; they preserve it with their charms.

We know we are favoured. We are strong, too, for we have fought many battles to safeguard our inheritance. We have crushed the skulls of our enemies with the bones of our strongest animals. We have repulsed all intruders. We have prevailed.

But still our enemies swarm about us with greed and murder in their eyes. They have ventured far from lands where rivers have dried and caked. They covet our land but we will not give in to them. We will die before then. But we need a strong leader. Who will be called to lead us now? A person marked by fortune for this great venture. One who will lead the clan to safety and victory. Our need is urgent; it is now. A leader must come forth in this time of peril.

✪ What kind of leader does this clan need in this time of crisis? Compose your own profile of the most suitable person to lead the clan.

As the clan's enemies draw near, various individuals present themselves as possible leaders to deal with this life-threatening situation. The frightened inhabitants huddle around closely to hear what they say.

A veteran of many battles speaks first. His delivery is loud and forceful:

I am **YOUR HERO**. I have saved you from disaster in the past and I'll do so again. I am the great warrior, the cunning hunter who fears not the most fearsome beast. Sabre-toothed tiger, woolly mammoth, marauding tribesman. You have seen me face them all unflinchingly and emerge victorious and unscathed. Follow me and have no fear. I will protect you as I always have.

And then a female member of the tribe steps forward. She speaks eloquently not of war but of persuasion.

LET ME PERSUADE THEM. You know I have the sweetest tongue in the land. You have heard me quieten the brute with my magical words. You have seen me stir the lazy into life and ease the wounded hunter into death. With my powers, I can entrance our enemies and persuade them to live alongside us – why suffer the agonies of war, when we can have a better, gentler future?

At this point, a loud voice barks from the shadows and a large aggressive man steps forward into the firelight.

Ignore the words of manipulative women. **I WILL FIGHT FOR YOU**. With my brute force I will break the bones of our enemies and scatter their crushed remains to the four winds. Might is right. The meek will perish. Only the strong flourish. Your cause is my cause. Let us unleash our anger. Let's unveil our awesome power. Stick, stone, axe and fire. Let's annihilate anyone who stands in our way.

Yet another steps forward to speak. The huddled masses bow in awe as the priestess says:

I CALL ON THE POWER OF THE GODS. My mystical intervention is our only hope of salvation. You have seen the miracles that I perform. Haven't I healed you when you were sick? Haven't I cast out the evil spirits that haunt you on moonless nights? Do my potions not turn the wolf from your door and take away the viper's sting? With the gods on our side, who dares to defeat us?

Now several voices are raised in agitated discussion. They speak of many things: the need for more land, justice and restitution, revenge and retribution. But the moment for decision has come. The clan must select its leader.

✪ While keeping the threat of invasion in your mind, look

carefully at the four claims for political power. Which do you consider to be strongest? Using a scale of 1–4, where 1 represents the strongest claim, rate them in order of effectiveness. Supply reasons for your opinions.

✪ Now, imagine you are the leader of the invading force. Your slogan is **WE NEED MORE LAND**. Write down the speech that you will use to motivate the members of your clan.

A primary reason for living in a group is the protection that a social community could offer against the dangers of wild animals or hostile neighbours. We can probably assume that these prehistoric social clans were led by whoever offered the greatest degree of security. Either by consent or simply by taking power, the strongest, most skilful inhabitant seized a position of authority. But there are several other advantages to group existence, and the extent to which individuals may have been able to fulfil the opportunities offered by these advantages would have influenced their particular role and position within the community.

The pros and cons of group existence

Pros	Cons
Social living is advantageous for sexual reproduction. Instead of an isolated individual having to find a partner, societies provide stable structures.	The major disadvantage is that individual members are required to set aside some of their selfish instincts in order to fit in with the actions and decisions of the collective.
Living in a group is advantageous to the rearing of children, especially in the case of humans, where infants are vulnerable and development to maturity extends over a long period of time.	Work and food need to be shared. This can lead to jealousy and competition as to who is allowed first access to the group's resources or, more contentiously, who is allowed to mate with whom?
Groups are more efficient than individuals at gathering food or hunting animals. By acting together they can tackle larger prey or collect more vegetable matter.	At a basic level, some mechanism needs to be in place to determine how the individuals comprising the group should interact with each other.
Groups respond to danger more effectively than individuals. They can collectively counteract threat, and individuals in the group may have specialized tasks, such as keeping watch or acting as decoys.	Unless members of the group have some agreed-upon understanding of interpersonal roles, competition, conflict and the ultimate destruction of the social unit is likely. Individuals may not always agree with group decisions.

People who live in groups are more likely to be able to adapt to changes in their external environment because they are able to share expertise and to pass on knowledge from one generation to another. The collective wisdom of a society is an asset, which accumulates over time and with diverse experience.	Group living implies the segregation of one population from other similar populations. Where natural resources are scarce, it is necessary to limit the size of a community and to protect its resources. This is best achieved through territorial division, which may be accomplished in various ways ranging from mutual consensus to force.

Of course, intergroup contact is unavoidable and may even be desirable. It is likely that Neanderthal people used a wide range of strategies to regulate the internal structure of their communities and to negotiate the tricky divide between self-protection and intergroup interaction. From Neanderthal burial habits, we can assume that a degree of reverence for the integrity of the individual, coupled with some system of belief, played an important role in social regulation. It would be wrong to assume that technological advancement implies social or religious advancement. Anthropologists studying communities who have lived in relative isolation from modern technology have shown that these groups have highly complex social structures and elaborate mechanisms for limiting conflict. It would be a mistake to view human history as a steady progression to ever-higher degrees of civilization or to see the political structures of our own society as some sort of ideal achievement.

Human Nature

Changes in political structures are often related to prevailing ideas about human nature. Let's now look at a number of different ideas about what it means to be human.

In Part 4 of Jonathan Swift's famous satirical novel, *Gulliver's Travels* (1726), Lemuel Gulliver – the surgeon and ship's captain with an unfortunate talent for being shipwrecked or cast away – is abandoned by his mutinous crew in Houyhnhnms' (pronounced 'winim') Land. Desolate and alone, he sets out to find the inhabitants of this territory in the hope of throwing himself at their mercy.

I FELL INTO A beaten road, where I saw many tracks of human feet, and some of cows, but most of horses. At last I beheld several animals in a field, and one or two of the same kind sitting in trees. Their shape was very singular, and deformed, which a little discomposed me, so that I lay down behind a thicket to observe them better. Some of them coming forward near the place where I lay, gave me an opportunity of distinctly marking their form. Their heads and breasts were covered with thick hair, some frizzled and others lank; they had beards like goats, and a long ridge of hair down their backs, and the foreparts of their legs and feet, but the rest of their bodies were bare, so I might see their skins, which were of a brown buff colour. They had no tails, nor any hair at all on their buttocks, except about the anus; which, I presume, Nature had placed there to defend them as they sat on the ground; for this posture they used, as well as lying down, and often stood on their hind feet ... The females were not so large as the males; they had long lank hair on their heads, and only a sort of down on the rest of their bodies, except about the anus, and pudenda. Their dugs hung between their fore-feet, and often reached almost to the ground as they walked. The hair of both sexes was several colours, brown, red, black and yellow. Upon the whole, I never beheld in all my travels so disagreeable an animal, nor one against which I naturally conceived so strong an antipathy. So that thinking I had seen enough, full of contempt and aversion, I got up and pursued the beaten road, hoping that it might direct me to the cabin of some Indian. I had not gone far when I met one of these creatures full in my way, and coming up directly to me. The ugly monster, when he saw me, distorted several ways every feature of his visage, and stared as at an object he had never seen before; then approaching nearer, lifted up his forepaw, whether out of curiosity or mischief, I could not tell. But I drew my hanger, and gave him a good blow with the flat side of it; for I durst not strike him with the edge, fearing the inhabitants might be provoked against me, if they

discomposed: upset
thicket: cluster of bushes

pudenda: external sex organs
dugs: breasts

antipathy: dislike

visage: face

hanger: sword

should know, that I had killed or maimed any of their cattle. When the beast felt the smart, he drew back, and roared
so loud, that a herd of at least forty came flocking about me

from the next field, howling and making odious faces; but I
ran to the body of a tree, and leaning my back against it,
kept them off, waving my hanger. Several of this cursed
brood getting hold of the branches behind leaped up into
the tree, from whence they began to discharge their excre-
ments on my head ...

In the midst of this distress, I observed them all to run
away on a sudden as fast as they could, at which I ventured
to leave the tree, and pursue the road, wondering what it
was that could put them into this fright. But looking on my
left hand, I saw a horse walking softly in the field, which my
persecutors having sooner discovered, was the cause of
their flight. The horse started a little when he came nearer
me, but soon recovered himself, looked full in my face with
manifest tokens of wonder: he viewed my hands and feet,
walking round me several times. I would have pursued my
journey, but he placed himself directly in the way, yet looking with a very mild aspect, never offering the least violence. We stood gazing at each other for some time; at last I
took the boldness to reach my hand towards his neck, with
a design to stroke it ... But this animal, seeming to receive

my civilities with disdain, shook his head, and bent his
brows, softly raising his left fore-foot to remove my hand.
Then he neighed three or four times, but in so different a

cadence: the rise and fall of
the speaking voice; pitch or
tone

cadence, that I almost began to think that he was speaking
to himself in some language of his own.

While he and I were thus employed, another horse came
up; who applying himself to the first in a very formal manner, they gently struck each other's right hoof before, neigh-
ing several times by turns, and varying the sound, which
seemed to be almost articulate. They went some paces off,
as if it were to confer together, walking side by side, back-
ward and forward, like persons deliberating upon some
affair of weight, but often turning their eyes towards me, as

it were to watch that I might not escape. I was amazed to see such actions and behaviour in brute beasts, and concluded with myself, that if the inhabitants of the country were endued with a proportionable degree of reason, they must needs be the wisest people upon earth.

proportionable: similar

(FROM: J. Swift, 1975 [1726]. *Gulliver's Travels*, pp. 269-271. Harmondsworth: Penguin.)

The repulsive beasts that Gulliver encounters are called Yahoos, while the gentle horses that he admires so greatly are the Houyhnhnms. Not only are Houyhnhnms the dominant species in this land, but they also possess a trait that is supposed to be present only in humans – reason. Although they are horses, they are rational creatures who live their lives entirely in terms of their intellectual capacity to know right from wrong and make choices solely on the basis of what is reasonable. It is the ability to reason that separates the Houyhnhnms from the Yahoos, who live by instinct alone. Whereas Houyhnhnm society is civilized and orderly, based on a strong sense of good manners and polite behaviour, the Yahoos exist in a state of anarchy. Gulliver describes them as 'cunning, malicious, treacherous and revengeful

JONATHAN SWIFT (1667–1745)

Jonathan Swift, born and educated in Ireland, is regarded as one of the greatest English satirists. His earliest prose work was *The Battle of the Books* (1697), which mocked the controversy then raging in literary circles over the relative merits of ancient and modern writers. This was followed by *Tale of a Tub* (1704), which cuttingly satirized the artificial pretensions of the literary and artistic élite. One of Swift's best-known satirical pieces, *A Modest Proposal* (1729), which he wrote in Dublin, made him a hero of the Irish people. This essay's attack on English policy towards Ireland revolves around the gruesome suggestion that the children of the Irish poor be sold as food to the wealthy.

Swift's masterpiece, *Gulliver's Travels*, was published anonymously in 1726. It was an instant success. Although the satire was originally intended as an allegorical and scathing attack on the vanity and hypocrisy of contemporary courts, statesmen, and political parties, it is also a damning reflection on human society as a whole.

... of a cowardly spirit, and by consequence insolent, abject, and cruel' (1975: 314). But Gulliver's shock is not limited to meeting these odd creatures. To his horror, the Houyhnhnms recognize him as a Yahoo, albeit a slightly more advanced form of the species. And when he encounters a Yahoo female while bathing in a river, she has no doubt that he would make a good mate. It is only the speedy intervention of his Houyhnhnm protector that saves Gulliver from what would for him be a fate worse than death! The similarities between humans and Yahoos are impossible to deny. Gulliver reluctantly acknowledges their likeness and comes to worship the Houyhnhnms so strongly that when he is forcibly evicted from their land (because he is seen as potentially dangerous) and returns home, he rejects his family and fellow men in favour of his horses.

The reader suspects the links between human and Yahoo from the first meeting. By establishing this similarity, Swift strikes his first satirical blow. Eighteenth-century England prided itself on being 'enlightened'. The Enlightenment was supposedly an age in which science and the arts flourished. Society was thought to be reaching the pinnacle of civilization. This was a time of confidence; a time in which the virtues of truth and reason could extinguish the darkness of a less harmonious past. More and more, people tended to think of humankind as essentially good. Swift, by contrast, is sceptical. In presenting the Yahoos as human equivalents, he relies on an older and infinitely more negative view of human nature; one that was perhaps best expressed in the middle of the seventeenth century by Thomas Hobbes (1588–1679).

THOMAS HOBBES

Hobbes was born in Malmesbury, England, and studied at Oxford University. In 1637, he became interested in the constitutional struggle between King Charles I and the English parliament, which later led to civil war and the execution of Charles in 1649. Hobbes's sympathies were with the king and in 1640 he privately circulated a defence of the king titled *The Elements of Law, Natural and Politic*. Fearing arrest by parliament, Hobbes fled to France, where he lived in voluntary exile. While living in Paris he wrote *Leviathan; or The Matter, Form and Power of a Commonwealth Ecclesiastical and Civil* (1651).

Leviathan: a monster of the sea; a thing of enormous size and power

Leviathan argues for a highly authoritarian political system, in which citizens, to secure their well-being, transfer their right of self-protection to an adequately powerful sovereign. This soereign may either be an individual or an assembly of people who has absolute and unquestionable power. Hobbes's belief in the need for such centralized power stems from his negative view of human nature. For Hobbes, people are governed by passion rather than reason. In their natural state, they are inherently competitive and greedy. More significantly, they live in continual suspicion and fear of one another. According to Hobbes, without the constraints of strong government, human life is characterized by spiritual and material poverty, chronic fear, insecurity and ignorance. He writes:

> ... [I]t is manifest, that during the time men live without a common Power to keep them in awe, they are in that condition which is called War ... Whatsoever therefore is consequent to a time of War, where every man is Enemy to every man; the same is consequent to the time, wherein men live without other security, than what their own strength, and their own invention shall furnish them withall. In such condition, there is no place for Industry; because the fruit thereof is uncertain; and consequently no Culture of the Earth, no Navigation, nor use of the commodities that may be imported by Sea; no commodious Buildings; no Instruments of moving, and removing such things as require much force; no knowledge of the face of the Earth; no account of Time; no Arts; no Letters; no Society; and which is worst of all, continual fear, and danger of violent death; And the life of man, solitary, poor, nasty, brutish, and short.
>
> (FROM: Thomas Hobbes, 1987. *Leviathan*, Part 1, Chapter 13, pp. 64–65. London: Dent. [spelling modernized])

is consequent to: is a result of

withall: with

commodious: comfortable

✪ Do you think Hobbes's view of human nature is accurate? Explain.
✪ To what extent do you think a powerful ruler or government is necessary to secure peace and security?
✪ Should the powers of a government be limited? If so, in what ways?

✪ What dangers might arise from granting a ruler or government absolute power?

Hobbes's views were controversial even during his own time. In 1666, his work was investigated for atheistic tendencies after a bill was passed by the House of Commons (the seat of the English government). Shortly after his death his works were destroyed by his own university, Oxford, in a bonfire of publications thought to be pernicious or harmful. Nevertheless, his arguments for an authoritarian state are not unique. They have formed a central theme in the works of earlier and later philosophers and thinkers, many of whom differ in significant ways from Hobbes's bleak outlook.

JOHN LOCKE

John Locke (1632–1704), a contemporary of Hobbes, also advocated strong and stable government. In *Essay Concerning Human Understanding* (1690) he argues that the mind of a person at birth is a *tabula rasa* (i.e., a blank slate), upon which experience imprints knowledge. Because of their inherent ignorance, all people live in intolerable anarchy when they are in a natural state without government. Unlike Hobbes, though, Locke believed that all persons are born good, independent and equal. Through science and logic (empirical thought), Locke argues, humans are able to overcome their natural ignorance and pursue their proper business, which is to seek happiness and to avoid misery. It is through logic, reason and the desire to secure a life of order and peace that people consent to be ruled.

These propositions were spelt out in greater detail in Locke's *Two Treatises of Government* (1690), where he attacked the divine right of kings and the type of all-powerful state conceived by Hobbes. In brief, Locke argued that power should not reside in the State but with the people. The State is supreme, but only if it is bound by civil and what he called 'natural' law. Sovereigns are entrusted with powers over their people in a social contract between ruler and citizens. According to Locke, if a government violates this contract by trying to steal people's private property, or by attempting to reduce them to slavery under arbitrary powers, it places itself in a state of war with the people, who are then not only absolved from further obedience, but also obliged to rise against the perpetrator.

A PROFILE OF JEAN JACQUES ROUSSEAU

Rousseau was notorious for his quarrelsome, sentimental, impulsive and unorthodox ways. Although he had little knowledge of music, he undertook to revolutionize musical notation. He had boundless energy and wrote operas, plays, novels, essays, political tracts, autobiography and social discourses. Several years of his life were spent in exile and his belief in the 'natural man', unspoiled by civilization, made him an idol of the French Revolution, which followed ten years after his death.

JEAN JACQUES ROUSSEAU

The French philosopher Jean Jacques Rousseau (1712–1778) had views on human nature that differed significantly from both Locke and Hobbes. *Discourse on the Origin of Inequality Among Mankind* (1755) expresses the view that science, art, and social institutions corrupt humankind and that the natural, or 'savage', state is morally superior to the civilized state. Rosseau continues this theme in his influential novel *Émile* (1762), which begins with the words, 'God makes all things good; man meddles with them and they become evil'. He develops similar ideas in his most famous political piece, *The Social Contract* (1762), where he writes, 'Man was born free, and everywhere he is in chains'. Given Rousseau's belief in the goodness of humans, the problem he confronts is how to devise a social system that will unite people and protect the person and goods of each member while retaining maximum individual freedom. His answer is the creation of a moral and collective body, brought into existence by general agreement of the whole community, to protect and promote the general will. In short, Rousseau calls for a system in which power resides in the community as a whole. Because this government owes its existence to the people who have made it, the people have a right to remove its power should it fail to serve their will.

'... the savage lives within himself, while social man lives constantly outside himself.'

JEAN JACQUES
ROUSSEAU

❂ In your view, what are the strengths and weaknesses of Locke's and Rousseau's beliefs on human nature?

❂ Do you believe that people have the right to remove a government from power, if necessary by violent means? Explain.

NATURE VS. REASON

At this point, we need to go back to Gulliver and the Houyhnhnms. It is clear that the Yahoos correspond with a Hobbesian view of humankind in its natural state. But, Gulliver, as the Houyhnhnms recognize, is no ordinary Yahoo. In their eyes he is a Yahoo with reason, and that's a dangerous thing to be. This is clear when Gulliver naïvely boasts of the enormous accomplishments and power of England. Here he describes their military prowess:

> ... cannons, culverins, muskets, carabines, pistols, bullets, powder, swords, bayonets, battles, sieges, retreats, attacks, undermines, countermines, bombardments, sea-fights; ships sunk with a thousand men, twenty thousand killed on each side; dying groans, limbs flying in the air, smoke, noise, confusion, trampling to death under horses' feet; flight, pursuit, victory; fields strewn with carcasses left for food to dogs, and wolves, and birds of prey; plundering, stripping, ravishing, burning, and destroying. (1975: 294)

● DO YOU THINK human nature is naturally good, bad or merely neutral (neither good nor bad)?

The Houyhnhnm master reacts with revulsion. To him, humans are animals at their most vile. If they possess reason, as they claim to do, then, he concludes, the corruption of reason may be worse than brutality itself. Instead of elevating society towards perfection, reason seems to allow humans to implement their abominable behaviour more effectively. Swift's message is clear. Human society may believe that it is based on rational, civilized values, but its actions constantly contradict this. Furthermore, while animals may be excused from 'savage' or self-interested behaviour because they do not have reason, similar behaviour in humans who have the faculty of reason is doubly damnable.

There is another subtle twist to Swift's cutting attack. The rational Houyhnhnms live in a state of nature. Incorruptibility and perfection characterize their lives. The name 'Houyhnhnm' means both 'horse' and 'the perfection of Nature' (1975: 281). Even their language has no word for lying or deceit; instead, when they wish to express disbelief in the incredible stories that Gulliver tells them, they have to refer to '[saying] the thing that was not' (1975: 281). Yet, in their pursuit of rational perfection, they show several disturbing signs. For example, they do not feel any deep affection for their colts or foals. The care that is given to offspring is dictated solely by reason. Nor do they have deep attachments to their partners. Relationships serve only to produce offspring: one of each sex. Moreover, relationships are not based on love, but on the physical characteristics that will produce the strongest colt or the most attractive foal. Their lives are austere, ordered and devoid of excitement or disagreement. Emotionless and immovable, theirs is a sterile and monotonous existence. Thus, at the heart of Swift's scathing satire is an attack on the idea that rationality is an adequate basis upon which to make decisions or to model personal conduct and social organization. This is Swift's challenge to (if not his rejection of) the smug self-satisfaction of his age.

- DO YOU THINK that human society should be governed solely on the basis of reason?

 Exercising Power

At this beginning of this section on 'Government and Governing', we took an imaginary journey to a Neanderthal settlement in order to look at why people may have formed themselves into societies and how power relationships may have developed. Subsequently, we looked at contrasting views of human nature, which in turn may have justified or promoted different types of government. It is easy to support a government in which a powerful monarch or a small élite group has absolute power if you regard humans as intrinsically ignorant, uncontrolled and dangerous. If, by contrast, you think human nature is inherently good and constructive, there is no need for excessive authoritarian control, and ideas of co-operative government appear both reasonable and necessary.

Of course, in life and in society things do not fall into place so easily and it would be quite wrong to see any direct link between philosophical claims and political institutions. At best, the arguments of philosophers found their way into the hearts and minds of a wider population. By provoking thought and challenging old truths with new possibilities, they created environments in which, when circumstances presented themselves, new forms of government took root. Once entrenched, power relations are extremely difficult to change. Those who hold the most power are reluctant to give it up. Those holding the least power often do not have the means to launch a successful challenge. Political change is usually measured in decades, sometimes centuries. Yet, in some instances, nations have been rocked by rapid, cataclysmic change. They have undergone revolutions.

The Oxford Advanced Learner's Dictionary defines 'revolution' in this way:

> **revolution** /ˌrevəˈluːʃn/ *n* **1** [C, U] an attempt to change the system of government, esp by force: *the French Revolution of 1789* ○ *a peaceful/bloodless revolution* ○ *stir up revolution.* **2** [C] ~ (**in sth**) a complete or dramatic change of method, conditions, etc: *a revolution in printing techniques/sexual attitudes* ○ *a technological/cultural revolution.* **3** [C, U] ~ (**on/round sth**) a movement in a circle round a point, esp of one planet round another: *the revolution of the earth on its axis round the sun* ○ *make/describe a full revolution.* See also COUNTER-REVOLUTION, INDUSTRIAL REVOLUTION.

A revolution therefore involves complete change: it is the overthrow of the political system in place, often by dramatic or violent means.

In the West, there have been four notable revolutions: the English Revolution (1640–1660), the American Revolution (1775–1783), the French Revolution (1789–1799) and the Russian Revolution (1917–1918). Although these revolutions differ in circumstance, with the exception of the American Revolution, they all involved the overthrow of a monarchy and its replacement with a collective government, more representative of the general population but not necessarily democratically elected. In the case of America, the revolution was a war of independence directed at colonial rule by Britain.

KINGS AND QUEENS: MONARCHIES

A monarchy is usually a form of government in which one person has the hereditary right to rule as head of state during his or her lifetime. Throughout history many monarchs have wielded absolute power, sometimes based on their claims to divinity. By the Middle Ages the monarchical system of government had spread all over Europe, where it was often based on the need for a strong ruler who could raise and command military forces to defend the country. European monarchies were dynastic, with the throne usually being passed on to the eldest son or nearest male descendant. Initially, many medieval rulers obtained soldiers and weapons from the feudal lords and thus were dependent on the support of the nobility to keep their power. With the decline of feudalism and the rise of nation-states, power became centralized in the hands of the sovereigns. At first, these rulers were supported by the growing middle class, or bourgeoisie, who benefited from a strong central government that kept order and provided a stable atmosphere in which trade could flourish. By the fifteenth and sixteenth centuries absolute monarchs ruled the countries of Europe.

A cornerstone of the power of these monarchs was the idea that sovereigns are representatives of God and derive their right to rule directly from God. This doctrine is known as the divine right of kings. According to this, a ruler's power is not subject to secular or worldly limitation; the ruler is responsible only to God. The seeds of the English Revolution were sowed when Charles I, claiming divine right, refused to give in to parliament. Popular support for ideas of the monarch's divine right to rule had begun to fade even before Charles I was beheaded, and with

- HOW MUCH
FREEDOM do
people really have,
even under demo-
cratic govern-
ment?

- HOW MUCH
FREEDOM should
they have?

the execution of Louis XVI and Marie Antoinette during the French Revolution they disappeared almost completely. Monarchs that have survived through the nineteenth and twentieth centuries have all been constitutional figureheads, with little real power to influence daily affairs. Over the last two hundred years the idea has taken hold that government should be voted in by the citizens of a country to serve their needs. Today, democratic rule, in which all adult citizens are able to participate, is the preferred form of government, even though there are still many exceptions. But this is a comparatively recent development. Who should vote and what rights ordinary citizens should have in relation to the state were (and is some cases still are) hotly contested issues.

Philosophers writing at the time of the French and American revolutions were divided on this point. Some, like Edmund Burke (1729–797), vehemently criticized the events in France and asserted the rights of monarchs. Others, like Thomas Paine (1737–1809), promoted the new ideals of liberty, equality and fraternity. Throughout his life, Paine was a champion of freedom. (One of his earliest articles called for the end of slavery.) But his reputation rests on two immensely important works.

TOWARDS DEMOCRACY

In January 1776, while living in Philadelphia, Paine anonymously published a 50-page pamphlet titled *Common Sense*. Using plain language and a forceful style, the pamphlet dramatically claimed that the American colonies received no advantage from Great Britain, which was exploiting them, and that common sense called for the colonies to become independent and establish a republican government of their own. *Common Sense* spurred the American colonists on to greater efforts in their war against Britain and greatly influenced the Declaration of Independence in 1776. Later, after spending several months in revolutionary Paris, Paine published *The Rights of Man*, in 1791 (Part One) and 1792 (Part Two). Not only did this work support revolutionary action and propose a republican form of government, it also included the famous Declaration of the Rights of Man and of Citizens, which had been adopted by the National Assembly of France in 1789. The first four clauses of the French declaration are repeated several times in Paine's work:

First: Men are born, and always continue, free and equal in respect of their rights. Civil distinctions, therefore, can be founded only on public utility.	**Second:** The end of all political associations is the preservation and imprescriptible rights of man. These rights are Liberty, Property, Security, and the right to resist oppression.	**Third:** The Nation is essentially the source of all sovereignty; nor can any individual, or any body of men, be entitled to any authority that is not expressly derived from it.	**Fourth:** Political Liberty consists in the power of doing whatever does not injure another. The exercise of the natural rights of every man has no other limits than those that are necessary to secure every other man the free exercise of the same rights. And those limits are determinable only by the law.

TIME TO DISCUSS

These claims of inalienable or absolute human rights have exerted an enormous influence on politics and society. We need to look at them carefully. Sit down with a friend and talk about the following issues:

✪ What does it mean to claim that people are born 'free and equal'? Does this suggest that everyone should have exactly the same claim to the wealth and benefits that a society may provide? Or does it simply imply that all people should have the same opportunities?

✪ What is meant by 'civil distinctions' and what does 'public utility' mean? If these words refer to differences in status among people, don't they contradict the ideal of equality?

✪ If the purpose of government is to preserve the rights of every person to liberty, property and security, can any government legitimately act to limit the individual? For instance, can any government founded on these principles pass legislation enforcing compulsory schooling for children, or banning a newspaper from publishing information, or preventing its citizens from owning guns?

- ✪ What is oppression? Does the 'right to resist oppression' include an armed struggle against the State, even if the government has been democratically elected?
- ✪ If the third clause suggests that power is located within the population as a whole rather than in any person or institution, how can it be implemented in any practical way? Isn't it necessary for people to grant the government the right to act on their behalf? If so, would it be permissible for a government to declare war on another nation without directly consulting the general population, for instance by holding a referendum?
- ✪ If all the citizens of a country agree to a policy that is potentially dangerous, say to the environment or to the inhabitants of another country, is it legitimate to pursue that policy?
- ✪ What are the implications of the fourth clause? Does this clause mean that we have to condone all behaviour as long as it does not infringe on the rights of other people? Does this freedom extend to attitudes and actions that religious belief may condemn as morally wrong, but that may not infringe the rights of any of the individuals involved?
- ✪ In your opinion, based on the rights outlined above, do you think that it is legal and right to use the death penalty as a punishment or to allow the abortion of a human foetus?

These are contentious issues and none of them has an easy resolution. Unless people discuss them and put their understanding into action, the founding clauses of Paine's work remain empty ideals – mere words on paper. But while many take these four clauses as self-evident truths of our basic rights as humans, other thinkers have questioned whether people are really free, even when they live in a democracy. Such a thinker was Karl Marx, and his writings had a direct bearing on the Russian Revolution and the system of government that followed it.

THE MARXIST IDEAL

Karl Marx (1818–1883) was born in Germany and educated at the universities of Bonn, Berlin and Jena. From early in his life his criticism of contemporary political and social conditions embroiled him in controversy. For Marx, capitalism was the central cause of human misery and the greatest impediment to political freedom for the masses.

His greatest works, *The Communist Manifesto* (1848) and *Das Kapital* (published as three separate volumes in 1867, 1885 and 1894), both of which were written together with Friedrich Engels, identify class distinction as the structure that underpins society and class struggle as the force that drives history. For Marx, the modern world is characterized by confrontation between the ruling capitalist class (the bourgeoisie) and the downtrodden working class (the proletariat). Each class is united within itself by the fact that it has a common economic interest. The capitalist bourgeoisie are united by the desire to generate further wealth by increasing production and by getting as much work out of their workers for as little pay as possible. The proletariat is united by a desire to increase their earnings and to diminish their exploitation by their capitalist employers. Divergent economic interests mean that each group is constantly in conflict with the other. Because the wealth of the capitalists is dependent on the sale of surplus production, it is essential that the means of production are kept out of the hands of the workers. Paradoxically, because capitalism seeks ever greater profit, it is constantly in search of more efficient economic production. This implies mechanization and the centralization of business – both of which ensure that the means of production and its associated wealth become concentrated in the hands of a smaller and smaller élite as less competitive concerns are driven out of business. At the same time, though, the proletariat increases in number and political awareness. Knowledge of exploitation and hatred of the monopoly of production, worsened by fluctuations in the economy, inevitably lead to growing popular discontent and to a revolution in which the bourgeoisie will be overthrown.

'Great ideas also have their fates.'

KARL MARX

According to Marx, it is only through a revolution, in which the ownership of private property is abolished and the means of production is taken from the capitalists and passed to the State (that is, the proletariat organized as the ruling class) that true freedom and equality can be achieved. Without changing the economic structure of society, all other changes are cosmetic window-dressing. This is because the ideas that prevail in any society are always those of the ruling class. In capitalist societies, the ideas that dominate are those of the capitalists themselves and they serve the interests of capitalism. These ideas, whether they are political, religious, economic or artistic, portray the structure of society as natural. In fact, the ideas of the ruling class represent false consciousness or ideology. They are designed to lull the general population into an acceptance of the nature and status of their lives. Ideology is insidious because it is ingrained into general consciousness and accepted unquestioningly. Ideology is a trap that ensures that the oppressed masses actively and unknowingly participate in their own oppression.

Even though some of the communist governments of the twentieth century have fallen, Marx's ideals have retained value, especially as huge numbers of people still live in abject poverty and the gap between the wealthy nations and the poorest nations of the world continues to grow. For many of the earth's inhabitants, freedom is an empty term as they face hunger and disease.

TIME TO WRITE

✪ Try to answer the following questions by using a library, an encyclopaedia and/or the Internet:
  – What is the gross national product (the total earnings generated) of the wealthiest nation on earth?
  – What is the gross national product of the poorest nation on earth?
  – How many people live below the poverty datum line?
  – How many children starve to death each year?

- How many people are illiterate?
- How many people in your country are unemployed?
- What percentage of the population of your country could be considered to be rich?
- What percentage of the population could be considered poor?

✪ Now imagine that you are a magazine journalist who has been commissioned to write an article on human rights. You're particularly interested in the enormous gap between wealth and poverty and you wish to alert your readers to this problem. The editor of your magazine thinks that this is a great idea, especially if you can make the article relevant to your context and if you can suggest ways of tackling inequality. Write an article that addresses these issues.

. . . .

SO FAR THIS chapter has focused on the social and political forms of power that we are most aware of in our daily lives: systems of government. To most of us, the words 'power' and 'politics' are synonymous with various governments and with the issues arising out of our experiences of being governed. Although the chapter has shied away from talking about the policies of any particular government or party, the discussion may have given the impression that power is something that happens to us: that we feel the effects of power without personally taking part in it. Of course, we might vote in an election or actively support a political party, but somehow this is not 'real' power. 'Real' power is something that is held by an élite few who sit in parliament and pass laws that the rest of us follow. In this way we can claim innocence when our government fails to wipe out corruption and tackle crime or when it remains silent about human rights violations in our own country or elsewhere. (As commentators have cynically noted, it is difficult to find anyone in South Africa today who supported the apartheid policies of the

THREE

LANGUAGE,
KNOWLEDGE AND
POWER

Nationalist government, even though this party was returned to power with increasing majorities in every election from 1948 to 1990.) Yet, whether we are conscious of it or not, power is inescapable. Individuals are never innocent of the exercise of power. It is as much a part of life as breathing or eating and it is just as involuntary. In this section, we will be looking at how power pervades our lives by examining the relationship between language, knowledge and power.

⬤ Language Matters

Early in her book, *Talking Power* (1990), Robin Lakoff makes a rather surprising claim. She writes: 'Language is powerful; language is power. Language is a change-creating force and therefore to be feared and used, if at all, with great care, not unlike fire' (1990: 13). What does she mean? How is language power? Why should we fear it and use it with great care?

To think about Lakoff's claim, try to answer the following questions:

✪ Two candidates are running for the position of chairperson of the students' Representative Council.

Candidate A has good ideas and sound policies. Unfortunately he is a poor communicator. He speaks hesitantly and often loses track of what he is saying. In public, he seems uncomfortable – seldom making eye contact. His manifesto is a complex document. It spells out an ambitious programme of action, but it is difficult to follow and has several grammar and spelling mistakes.

Candidate B also seems to be competent, although her vision for the Council has limitations and her policies are a little controversial. She is a powerful speaker. Not only can she debate aggressively, but she also has the ability to motivate people to take action. Her self-confidence shows in her contact with people and in her manifesto, which is clear, grammatically flawless and to-the-point.

Would you vote for Candidate A or Candidate B? Why? If you voted for Candidate A, do you think he would win the election?

✪ Imagine that you have been arrested for a serious crime for which the penalty is long-term imprisonment. You're innocent of the charges, but the police have forensic evidence that places you at the scene of the crime and they have the testimony of witnesses that suggests that you had a motive for the crime. You need to choose a lawyer to defend you in court. Make a list of the characteristics you would look for.

Lawyers and politicians, as Lakoff points out, are particularly dependent on language. Their success or failure often depends on how well they are able to use language to persuade other people to adopt their views. A good lawyer needs to know how to highlight information that is favourable to a client and how to downplay evidence that suggests his or her guilt. In a courtroom battle, truth does not necessarily win the day, but the version of truth that is most convincing. Nor do intricate policies on taxation or foreign policy ensure victory for the politician. Rather, if you want to be president you need to have (or appear to have) charisma, energy, drive, and the ability to outwit your opponents. And language plays a crucial role in all political operations. In Lakoff's words:

> Language drives politics and determines the success of political machinations. Language is the initiator and interpreter of power relations. Politics is language ... At the same time, language is politics. How well language is used translates directly into how well one's needs are met, into success or failure, climbing to the top of the hierarchy or settling around the bottom, into good or bad relationships, intimate and distant. Language allocates power through politics, defines and determines it, decides its efficacy. (1990: 12–13)

It is the way that language translates into power that makes it dangerous. Language is a manipulative and persuasive force. It has the capacity to influence and change human behaviour and society for good or for bad. Language is also a defining force. When we label ideas or people or cultural groups in particular ways, we lay the foundations for our treatment of them. By calling our opponents 'fanatics', 'fools' or 'dictators', we justify our behaviour towards them. Similarly, labels like 'retarded', 'crippled' and 'queer' have been used legally to strip individuals (who do not conform to society's notions of what is normal) of their human rights. Finally, language can be a mechanism for control. By determining what may or may not be said, it is possible to control what may be thought and done. Censorship of books, newspapers and films works on this basis. But in extreme (and all too frequent) instances, any critic of the government may be labelled a 'traitor' and any criticism 'treason'.

Given the potential effects of language, we might take comfort from the idea that as individuals we have never used language to damage anyone. And while it might be true that our words have not led directly to a massacre, linguists who have examined everyday interactions between people have noted power at play. We all manipulate language and we are all always involved in persuasion of some sort or another. As Lakoff shows, even apparently innocent conversational exchanges entail power. Let's look at an example:

> He: Do you feel like going to the movies?
> She: Mmm. I don't mind.
> He: What do you wanna see?
> She: Oh, I don't know ... you choose.

This is a traditional form of the male/female game. I'm sure that most of you are familiar with conversations like this. He has made a suggestion. She can either accept it or propose an alternative. Either position would involve self-assertion. She would have to take responsibility for her decision and the success or failure of the date would then rest on her shoulders. To avoid this, the decision is left up to him. But she never gets to do what she wants to do and even if they enjoy the movie, it has been his choice and he gets to take the credit for it. In this way, she retains the power that she has because she cannot be proven to have made a wrong choice. At the same time, because he always makes the decisions, she can complain that he always gets his way and thus she can retain a manipulative hold on him. Of course, he is also implicated in the power play. By asking her to make the decisions he creates the illusion of being open to her control, while knowing that the choice will be left up to him. Notice that he doesn't ask: 'Do you want to see *Rambo XVII*?' This may be more likely to elicit a direct response and perhaps deprive him of making the final decision.

Clearly, the conversation would have a different meaning or significance if the gender roles were reversed or if it was taking place between two people of the same gender.

✪ Analyse the power relations in the following conversation. Who has the advantage here?

> A: Are you busy on Saturday?
> B: No, why?
> A: Well, I'm having a party and I would like you to be the barman.

✪ In what ways does the following exchange entail the exercise of power?

> Patient: I have an appointment with Dr Da Silva at three thirty.
> Receptionist: Oh, Doctor's running late. Please have a seat.

✪ From memory, transcribe (write down) an argument that you have had with a partner or friend. (Please don't start an argument for the sake of this exercise!) Once you have done so, explain the way in which language reveals the changing relationships of power during the argument.

Language can manipulate in very subtle ways: it can depend on the choice of one word rather than another or on the manner in which a sentence is constructed. For example, it is common to use the passive voice to obscure the relationship between the subject of a sentence and the action in which the subject is engaged. The passive sentence 'Legislation has been enacted to prohibit unlawful strikes' is far more vague and evasive than its active version: 'Parliament (or the government) has enacted legislation to prohibit unlawful strikes.' Similarly, language can be used to suggest a particular position without explicitly stating it. One example, which appears in this book in various forms, is the construction 'Some people believe ...' or 'Some critics argue ...'. In these instances, the use of the indefinite pronoun 'some' could imply that this is a minority view and therefore that it is of limited value. So a textbook, which may seem to provide an objective analysis, is itself engaged in covertly promoting specific ideas. We will return to this matter a little later, but now are going to examine the strategies used by the true masters of linguistic manipulation – the makers of political speeches.

 ## Political Rhetoric

The people of the United States of America often claim to have the greatest democracy on earth. They have been working at refining their government since Thomas Jefferson drafted the Declaration of Independence in 1776. Yet few Americans would argue that their country is a utopia of political perfection. All are equal under the constitution, but many Americans know through experience that some are more privileged than others. Where privilege is entrenched, abuse of power inevitably invites resistance from those who consider themselves to be oppressed.

People know that they deserve better than outrage and exploitation. And writing is a powerful tool of rebellion. One of the most famous texts of resistance ever written about the imperfections of society is Martin Luther King's 'I Have a Dream' speech. King knew that democracy was theoretically in place in his country, but he also

WHO WAS MARTIN LUTHER KING?

Martin Luther King Jr. (1929–1968) was a US civil rights campaigner, black leader and Baptist minister.

Born in Atlanta, Georgia, son of a Baptist minister, King founded the Southern Leadership Conference in 1957. A brilliant and moving speaker, he was the symbol of, and leading figure in, the campaign for integration and equal rights in the late 1950s and early 1960s. He came first to national attention as leader of the Montgomery, Alabama, bus boycott in 1955, and was one of the organizers of the massive (200 000 people) march on Washington D.C. in 1963 to demand racial equality. An advocate of non-violence, he was awarded the Nobel Peace Prize in 1964. But in spite of international recognition, he was the target of intensive investigation by the federal authorities, chiefly the FBI under J. Edgar Hoover. He was assassinated in Memphis, Tennessee, by James Earl Ray, although there are abiding doubts about whether Ray acted alone or even whether he was responsible for the murder. King's birthday (15 January) is observed on the third Monday in January as a public holiday in the USA.

yearned for that true democracy where there would be equal opportunity for all. He wanted America to 'get it right'. On August 28 1963, Martin Luther King gave a speech to his black compatriots who had marched on Washington D.C., the capital of the United States, to protest against prejudice, their lack of human rights, and consequently, the assault on their dignity.

Let's look at an extract from King's 'I have a Dream' speech.

... I say to you today, my friends, that in spite of the difficulties and frustrations of the moment I still have a dream. It is a dream deeply rooted in the American dream.

I have a dream that one day this nation will rise up and live out the true meaning of its creed: 'We hold these truths to be self-evident; that all men are created equal.'

I have a dream that one day on the red hills of Georgia the sons of former slaves and the sons of former slave owners will be able to sit down together at the table of brotherhood.

I have a dream that even the state of Mississippi, a desert state sweltering with the heat of injustice and oppression, will be transformed into an oasis of freedom and justice.

I have a dream that my four little children will one day live in a nation where they will not be judged by the colour of their skin but by the content of their character.

I have a dream today.

I have a dream that the state of Alabama, whose governor's lips are presently dripping with the words of interposition and nullification, will be transformed into a situation where little black boys and black girls will be able to join hands with little white boys and white girls and walk together as sisters and brothers.

I have a dream today.

I have a dream that one day every valley shall be exalted, every hill and mountain shall be made low, the rough places will be made plain, and the crooked places will be

American dream: the traditional ideals of the American people, such as equality, democracy and material prosperity

self-evident: obvious, clear

interposition: interruptive words
nullification: denial; lies

jangling: making a harsh,
metallic sound

made straight, and the glory of the Lord shall be revealed, and all flesh shall see it together.

This is our hope. This is the faith with which I return to the South. With this faith we will be able to transform the jangling discords of our nation into a beautiful symphony of brotherhood. With this faith we will be able to work together, to pray together, to struggle together, to go the jail together, to stand up for freedom together, knowing that we will be free one day.

This will be the day when all of God's children will be able to sing with new meaning 'My country 'tis of thee, sweet land of liberty, of thee I sing. Land where my fathers died, land of the pilgrim's pride, from every mountainside, let freedom ring.'

prodigious: marvellous;
amazing or large;
enormous

And if America is to be a great nation this must come true. So let freedom ring from the prodigious hilltops of New Hampshire. Let freedom ring from the mighty mountains of New York. Let freedom ring from the heightening Alleghenies of Pennsylvania!

Let freedom ring from the snowcapped Rockies of Colorado!

Let freedom ring from the curvaceous peaks of California!

But not only that; let freedom ring from Stone Mountain of Georgia!

Let freedom ring from every hill and molehill of Mississippi. From every mountainside, let freedom ring.

When we let freedom ring, when we let it ring from every village and every hamlet, from every state and every city, we will be able to speed up that day when all of God's children, black men and white men, Jews and Gentiles, Protestants and Catholics, will be able to join hands and sing in the words of the old Negro spiritual, 'Free at last! Free at last! Thank God almighty, we are free at last!'

28 August 1963
Washington, D.C.

POWER

✪ Do you find this speech impressive? Why?

✪ If you had been a member of the 200 000-strong audience who listened to Martin Luther King, how do you think you would have felt at the end of the speech?

✪ Would you have been disappointed at the lack of a specific agenda? Would you have liked him to spell out a particular plan of action?

✪ Martin Luther King uses a lot of biblical imagery. In your opinion, does this strengthen or weaken his speech? If you take his largely religious audience into account does this make a difference to your perspective?

✪ Do you find the repetition of 'I have a dream' irritating, or does it gather more power with each restatement?

✪ What devices other than the repetition of 'I have a dream' does King use to stir his audience?

Political speeches are deliberately crafted to achieve their purpose. Politicians often use professional speech writers who spend many hours working on drafts of what is going to be said. These writers take into account the purpose of the speech, the audience to whom it will be delivered and the effect that they want to elicit. In shaping ordinary words into a highly persuasive form they may use some (or all) of the following strategies:

Emotive language If you say, for instance, that a former leader is 'a traitor with the blood of innocent people on his hands' you are using language full of resentment and anger. Strong verbs and many adjectives often indicate strong emotions. A 'strong verb' is a word that expresses a drastic action or intense feeling. 'Hate', 'kill', 'adore', 'rape' and 'fight' are all strong verbs.

Inclusivity This is when you address your audience as 'us'. This tactic includes your listeners, and makes them feel as if you are on their side.

Flattery People enjoy 'feeling good' about themselves. If they are told that they are intelligent, for instance, or on the right side, they are likely to warm to the speaker. Because good speakers 'read' their audiences, they respond by characterizing them in ways designed to please them. They also include what they believe their listeners want to hear.

Insult On the other hand, if their 'enemies' are insulted, they can become worked up against them. Hitler's insulting 'hate speech', aimed against the Jews, eventually led to the murder of six million people. It is said that 'the pen is mightier than the sword' but in some instances the tongue is stronger than both.

Repetition Skilled speakers have found that people often lose concentration, or reflect on particular points that have been made, thus missing some of the speech. They feel that it is a good idea, therefore, to repeat the main points of the address. Repetition also has a cumulative effect in that it can whip up emotion. Notice how Winston Churchill, Britain's leader during World War II, makes use of repetition (as well as inclusivity) in this famous speech to the House of Commons on 4 June 1940:

> We shall go on to the end, we shall fight in France, we shall fight on the seas and oceans, we shall fight with growing confidence and growing strength in the air, we shall defend our island, whatever the cost may be, we shall fight on the beaches, we shall fight on the landing grounds, we shall fight in the fields and on the streets, we shall fight in the hills; we shall never surrender ...

Climax Many speeches that are designed to move audiences emotionally build to a very strong climax. Churchill's speech climaxes in the stirring resolution 'We shall never surrender'.

Figurative Language People tend to remember pictures long after they have forgotten words. Many skilled speakers therefore use vivid images as an aid to their audience's memory. A famous example is Harold Macmillan's speech to the white South African parliament on 3 February 1960. The phrase 'the wind of change is blowing through the continent' is remembered by many.

Display of personal emotion Many skilled speakers and demagogues (political agitators appealing to the basest desires and prejudices of the mob) display personal emotion. This is sometimes sincere, sometimes an attempt to appear sincere, but most often to manipulate the audience. They raise their voices, alter the cadence (the rise and fall of the speaking voice) of their speech and make use of significant pauses to drive their message home.

Incitement to action After rousing crowds with emotive and manipulative language, speakers often finish speeches with calls for action.

You probably noticed that Martin Luther King makes use of several of these rhetorical devices in his speech. (Go back to your original analysis and see which ones you identified and which you missed.) The success of this speech owes much to the noble ideals that it expresses, but also to King's skill in manipulating language. King tackles the issue of racial inequality head-on and inspires his audience to continue with the struggle for freedom.

Such forthright expression from a political leader is often the exception. Yet without running down all politicians, we can note another important aspects of political persuasion, namely generalization. For an example we are going to stay in America and look at an extract from President John F. Kennedy's (1917–1963) inaugural speech, which was delivered in Washington on 20 January 1961:

> So let us begin anew – remembering on both sides that civility is not a sign of weakness, and sincerity is always subject to proof. Let us never negotiate out of fear. But let us never fear to negotiate.
>
> Let both sides explore what problems unite us instead of belabouring the problems that divide us.
>
> Let both sides join to invoke the wonders of science instead of its terrors. Together let us explore the stars, conquer the deserts, eradicate disease, tape the ocean depths, and encourage the arts and commerce.
>
> Let both sides unite to heed in all corners of the earth the command of Isaiah – to 'undo the heavy burdens ... and let the oppressed go free'.
>
> Now the trumpet summons us again – not as a call to bear arms, though arms we need – not as a call to battle, though embattled we are – but a call to bear the burden

● THE OLD JOKE asks 'How do you know when a politician is lying?' The reply is 'When he moves his lips.'

belabouring; making too much of

Isaiah: a biblical prophet

of a long twilight struggle, year in and year out, 'rejoicing in hope, patient in tribulation' – a struggle against the common enemies of man: tyranny, poverty, disease, and war itself.

Can we forge against these enemies a grand and global alliance, north and south, east and west, that can assure a more fruitful life for all mankind? Will you join in that historic effort?

In the long history of the world only a few generations have been granted the role of defending freedom in its hour of maximum danger. I do not shrink from this responsibility – I welcome it. I do not believe that any of us would exchange places with any other people or generation. The energy, the faith, and the devotion which we bring to this endeavour will light our country and all who serve it – and the glow from that fire can truly light the world.

And so, my fellow Americans: ask not what your country will do for you – ask what you can do for your country.

My fellow-citizens of the world: ask not what America will do for you but what together we can do for the freedom of man.

20 January 1961
Washington D.C.

endeavour: project, attempt

Kennedy had won the election by the narrowest of margins and at a time when tensions between the United States and the Soviet Union were at their height. To many, the world seemed poised on the brink of a new world war and electing a young man to the high office of president (at 43 he was the youngest president ever) was a dangerous move. Kennedy's aim in this speech is to address as wide an audience as possible. He wanted to win the support of his former political opponents and to unite the nation behind a single purpose. At the same time, he was anxious to make his mark as an international statesman. In order to do this, it was

POWER

necessary for the new president to make a speech that was both visionary and inclusive. It had to be crafted so that it would not alienate anyone. Therefore, his speech avoided any specific position or programme. While the speech sketches an appealing vision of the future, it is extremely vague about the policies that will be needed to achieve this. Generalization is a common device in many memorable political speeches. By steering away from specific details and by concentrating on abstract values such as freedom, justice and faith the speaker is able to tap into people's need to feel good about themselves and their country. Any statement that might cause offence is balanced by a corresponding statement that cancels out the initial assertion. The audience can and probably does select what they want to hear, so that the speech functions by speaking to individuals according to their personal interpretation of it. Because the speech is inclusive and generalized, each listener is able to construct it in terms of their preferences. Contrary to popular wisdom, which states that you can only please some people some of the time, it attempts to please *all* of the people *all* of the time.

TIME TO WRITE

To consolidate your understanding of how politicians manipulate language in their political speaking, answer the following questions on Kennedy's inaugural address in your journal:

✪ Make a list of the pronouns used in the speech. How do they change from the beginning of the speech to the end? What effect does this change have?

✪ Which key words or phrases are repeated in the speech? Why do you think the speaker chose to use these repetitions?

✪ The structure of many of the sentences in Kennedy's speech is balanced. The technical term for a balanced, parallel structure used to highlight contrasting ideas is 'antithesis'. Make a list of as many antithetical structures as you can find. What purpose do these structures serve?

● AN EXAMPLE of antithesis is Keats's famous phrase, 'truth is beauty, beauty truth'.

❂ Read the speech aloud. Which sounds are repeated and which sounds rhyme? In your opinion, does the use of sound add to or detract from the power of the speech? Why?

❂ How does the speech appeal to people's emotions? Pick out some emotive words and explain what effect they might have.

❂ Why do you think the president included a biblical reference in the speech? What do you think he wanted to convey about himself?

❂ What do you think of the final two paragraphs of the speech?

❂ Now, imagine that you have been elected president of your country. In a page or two, write down your inaugural address.

 Prescribing Language

At the beginning of this section on 'Language, knowledge and power', I said that language could be used as a mechanism for controlling people. To illustrate this point, we are going to look at a chilling extract from George Orwell's novel *Nineteen Eighty-Four*, which offered a bleak view of the future of society when it was first published in 1947.

Nineteen Eighty-Four is set in Oceania in the year of the title. The world is divided into three zones that are constantly at war with each other: Oceania (which includes what was known as Britain, the Americas, Australasia and Southern Africa), Eurasia (roughly Europe and Russia) and Eastasia (China, Japan, Manchuria, Mongolia and Tibet). The leader of Oceania is Big Brother, who appears only as an image on television screens and posters, and the system of government is highly authoritarian.

GEORGE ORWELL (1903–1950)

George Orwell, the pen name of Eric Arthur Blair, is best known for his political satires, both of which have sold over two million copies: *Animal Farm* (1945) and *Nineteen Eighty-Four* (1947). Many of his books, though, deal with social and political issues. *Down and Out in Paris and London* (1933) is an account of the appalling circumstances of the homeless poor. *Burmese Days* (1934) is a scathing indictment of imperialism. *Homage to Catalonia* (1938) is a moving account of the Spanish Civil War (1936–1939). And *The Road to Wigan Pier* (1937) is a harrowing report on the conditions of unemployed coal miners in the north of England. You might like to read one (or more) of these books.

Citizens are always under surveillance; the dominant slogan is 'BIG BROTHER IS WATCHING YOU'. The central character of the novel is Winston Smith, who works at the Ministry of Truth, which controls all information and which is in the process of compiling a new language, 'Newspeak'. We are going to join Winston as he has a conversation with Syme, who is employed by the Research Department at the Ministry of Truth:

'How is the Dictionary getting on?' said Winston, raising his voice to overcome the noise.

'Slowly,' said Syme. 'I'm on the adjectives. It's fascinating.'

He had brightened up immediately at the mention of Newspeak. He pushed his pannikin aside, took up his hunk of bread in one delicate hand and his cheese in the other, and leaned across the table so as to be able to speak without shouting.

'The Eleventh Edition is the definitive edition,' he said. 'We're getting the language into its final shape – the shape it's going to have when nobody speaks anything else. When we're finished with it, people like you will have to learn it all over again. You think, I dare say, that our chief job is inventing new words. But not a bit of it! We're destroying words – scores of them, hundreds of them, every day. We're cutting the language down to the bone. The Eleventh Edition won't contain a single word that will become obso-lete before the year 2050.'

He bit hungrily into his bread and swallowed a couple of mouthfuls, then continued speaking, with a sort of pedant's passion. His thin dark face had become animated, his eyes had lost their mocking expression and grown almost dreamy.

'It's a beautiful thing, the destruction of words. Of course the great wastage is in the verbs and adjectives, but there are hundreds of nouns that can be got rid of as well. It isn't only the synonyms; there are also the antonyms. After all, what justification is there for a word that is simply the opposite of some other word? A word contains its opposite in itself. Take 'good', for instance. If you have a word like 'good', what need is there for a word like 'bad'? 'Ungood' will do just as well – better, because it's an exact opposite, which the other is not. Or again, if you want a stronger version of 'good', what sense is there in having a whole string of vague useless words like 'excellent' and 'splendid' and all the rest of them? 'Plusgood' covers the meaning; or 'doubleplus-good' if you want something stronger still. Of course we use those forms already, but in the final version of Newspeak there'll be nothing else. In the end the whole notion of goodness and badness will be covered by only six words – in reality, only one word. Don't you see the beauty of that, Winston? It was B.B.'s idea originally, of course,' he added as an afterthought.

A sort of vapid eagerness flitted across Winston's face at the mention of Big Brother. Nevertheless Syme immediately detected a certain lack of enthusiasm.

'You haven't a real appreciation of Newspeak, Winston,' he said almost sadly. 'Even when you write it you're still thinking in Oldspeak. I've read some of those pieces that you write in *The Times* occasionally. They're good enough, but they're translations. In your heart you'd prefer to stick to Oldspeak, with all its vagueness and its useless shades of meaning. You don't grasp the beauty of the destruction of words. Do you know that Newspeak is the only language in the world whose vocabulary gets smaller every year?'

Winston did know that, of course. He smiled, sympathetically he hoped, not trusting himself to speak. Syme bit off another fragment of the dark-coloured bread, chewed it briefly, and went on:

'Don't you see that the whole aim of Newspeak is to narrow the range of thought? In the end we shall make thoughtcrime literally impossible, because there will be no words in which to express it. Every concept that can ever be needed, will be expressed by exactly one word, with its meaning rigidly defined and all its subsidiary meanings rubbed out and forgotten. Already, in the Eleventh Edition, we're not far from that point. But the process will still be continuing long after you and I are dead. Every year fewer and fewer words, and the range of consciousness always a little smaller. Even now, of course, there's no reason or excuse for committing thoughtcrime. It's merely a question of self-discipline, reality-control. But in the end there won't be any need for that. The Revolution will be complete when the language is perfect ...'

(FROM: George Orwell, 1979. *Nineteen Eighty-Four*, pp. 44–45. Harmondsworth: Penguin.)

TIME TO REFLECT

✪ In your opinion, are there any benefits in simplifying a language, especially a complex and sometimes grammatically irrational language like English?

✪ Do you think that words are always adequate for expressing ideas, feelings or things?

✪ What advantages does Syme see in 'destroying words'?

✪ What effect do you think Newspeak would have on literature?

✪ Do you think that by removing words it is possible to control what people think? Why?

✪ The South African constitution prohibits 'hate speech'. Any utterance that incites or promotes racial hatred or discrimination is illegal. Do you think that words really have the power to provoke people to act in undesirable ways? Do you think that a government has the right to decide on what can be said and what cannot?

✪ Can you think of any circumstances in which someone has tried to control your use of language? Jot these circumstances down.

Orwell's predictions for the year 1984 have not come about (or have they?). The major languages of the world are flourishing. New technology and the global revolution in communication have made this the information age.

Significantly, it is possible to trace the origins of modern political systems back to an earlier technological breakthrough – the invention of the printing press by Johann Gutenburg in 1450. Prior to this, books were copied by hand. The enormous labour involved made them extremely expensive. As a result, the knowledge they contained was confined to a small élite group of nobility and clergy. For the most part, the general population remained illiterate and dependent on the few people with access to books and the ability to read for their information. This gave enormous power to the Church and the ruling élite. They could control what people knew. The printing press drove down the cost of reproducing knowledge. With increased access to knowledge, people were able to acquire new skills and to change the material conditions of their lives. Slowly, these changes in knowledge and status led to demands for a greater say in the affairs of government. Language and knowledge, directly and indirectly, fuelled the fires of social and political change. Even today, political movements in various parts of the world recognize the importance of language. In Scotland, Ireland and Wales, for example, the revival of the ancient languages of the people has been at the centre of the drive for greater freedom from the dominance of the English parliament in London. Promoting a national language is inseparable from promoting a national culture and, consequently, a shared sense of identity.

You probably know that being able to speak the official language of a country fluently is an advantage. People who speak the official language have greater access to opportunities and resources. If you cannot speak, read and write the official language of a country, you are likely to find yourself excluded from both wealth and power.

South Africa has eleven official languages; but in reality English is the primary means of communication (the *lingua franca*) in the political and economic spheres. This places English speakers in a privileged position, but it also implies that they should be sensitive to the rights of speakers of other languages. English is a colonial language. It has been imposed on countries where it was not naturally spoken by the indigenous population. Moroever, its rise to prominence has not simply been a matter of coincidence. In fact, it has become South Africa's official language because of the interests and influence of a particular political group, who are mainly white and who, under the previous political ruling, severely limited the rights of other racial groups. Finally, the dominant position of English has overshadowed the richness and beauty of our African languages.

'OFFICIAL' LANGUAGE

- How many languages are spoken in your country?
- Which language is most commonly used by businesses and for official communication?
- Is this 'official' language the first or home language of most of the citizens?
- Do you speak the 'official' language of your country as your home language? If not, where and how is your home language used? How do you feel about this?

As we noted in Chapter two, when critics speak of South African literature they usually mean South African *English* literature.

Let's try to find out why by considering the following questions:

Which of the following seems to you to be an acceptable sentence in English?

✪ The students is writing an examination in the library.
✪ I want you to really try.
✪ Let's keep it a secret between you and I.
✪ That is the man I was talking to.
✪ I will meet you just now.

All of these sentences make sense. All of them are used by millions of people in some form or another every day. Even this book uses constructions that are similar to some of the sentences above. (The last sentence is a case in point.) Nevertheless, in terms of standard English, all of the sentences above are grammatically incorrect. (If you would like to know why, the reasons are printed upside-down on this page.) We need to be aware that grammar can be applied prescriptively to disempower people. By applying an inflexible standard, language purists deny that regional variations in usage may constitute perfectly valid communication. Deviations from prescriptive grammar are at worst considered to be signs of stupidity and incompetence. Variations from standard English (even shifts in pronunciation) are also sometimes seen as a sad indication of the poor state of the country and the decline in civilized values. Often these complaints, which are typically directed at radio and television announcers or at newspaper journalists, thinly conceal an extremely conservative political attitude.

English as a language comes with an entire culture that has its basis in social and value systems that come from England. It is simply impossible to separate language from culture. Each time a language is used, a particular set

Sentence (a) has a concord error. The subject and the verb don't agree in number; the writer should have used 'are' instead of 'is'. In sentence (b), the infinitive 'to try' is split. Sentence (c) uses the subject form of the pronoun rather than the object form. To correct it, we have to replace 'I' with 'me'. Sentence (d) ends on a preposition. Strictly speaking, this is unacceptable. Sentence (e) incorrectly uses the word 'just'. The primary meaning of 'just' is 'what is morally right or fair', although it can mean 'well deserved' (as in 'he received his just reward' and 'the exact or correct amount' (as in 'just enough'). The use of 'just now' to mean 'in a little while' is a peculiarity of South African English.

of values are articulated. With every utterance, these values seem more natural and more normal until we accept them without question, even though they may be, at best, inappropriate and at worst, harmful. Although South Africans may have tried to tame the English language and make it their own, a foreign cultural identity lurks beneath the surface. Inevitably, and often without our noticing, this foreign cultural identity exerts its power. As users of the language, we have to decide how we should react.

How does this foreign cultural identity assert itself and what does this have to do with power? Let's now look at the relation between education and language. After we have looked at this, you may want to think of other ways in which the control of language translates into power.

 Education and language

Our second example of how language and knowledge become instruments of power is also directly relevant to us, because, as writers of this book and as readers, we are involved in education. Schools, colleges and universities are the main instruments for socialization in contemporary society. Through these institutions people are taught how to become useful members of society. Their enormous influence is reason enough to examine them and their practices closely.

If you've read *Critical conditions* carefully, you will already be aware of how the choice of syllabus and the ways in which the syllabus is presented can promote particular views and attitudes while denying others. Academics are seldom asked to explain why they teach what they do or to justify their approaches. Instead, decisions about the content of a subject are kept hidden from the public at large. More importantly, the control over 'academic' decisions remains in the hands of academics, although even here there may be hidden influences. For example, any change to the teaching programme in a university needs the approval of fellow academics, both within the subject area and the wider university community. A syllabus change will go through various stages of scrutiny in the university hierarchy, until it receives the approval of the governing body. The initiating academic is not free of political ideas; nor are the various committees and forums that are consulted. Similarly, the institution as a whole may be susceptible to political influence, especially if it relies on government funding for its survival. Thus, rather than regarding school and university syllabuses as presenting objective versions of truth or reality, we need to treat them with suspicion. What version of truth or reality is really being offered? Does this version work to maintain the status quo or does it challenge the existing distribution of power in society?

There is another dimension to the link between educational institutions and social control that is even more basic than the choice of syllabus, and this dimension is language. Unlike any other institution, schools and universities trade in language. A business may use language in the process of producing computers, but the final product is a computer and the process of production is not intrinsically communicative. In schools and universities, however, language is both the product and the means by which it is produced. Moreover, language is central in defining the quality of the production: in this case, in separating those who have been educated successfully from those who fail. In order for learners to prove their worth, they need to be able to display their knowledge by showing that they have mastered the language of their discipline.

In itself, it is not negative to ask people to be able to use language. But because access to language and knowledge mean access to power, both are jealously guarded.

Particularly in universities, students are expected to learn how to write (and speak) in a very formal way. Academic writing has its own rules and conventions that set it apart from ordinary communication. Contrary to what might be expected, complexity is valued more highly than simplicity. The hidden logic goes something like this: if your writing is too easy to follow, it obviously lacks depth and substance. If everyone can grasp your discipline, then it has little value and brings even less status. Clearly, worthwhile subjects are those which are difficult to understand. Subjects that are difficult to understand can only be expressed in

JARGON

Jargon is the specialized language used in a particular discipline. Various disciplines have developed highly specialized languages, far removed from everyday use, to segregate those who may practise the discipline from those who may not. Think of the vocabulary used by medical doctors or geologists or literary critics. If you can't tell the difference between 'catalexis' and a 'choriamb', then you certainly cannot belong to the club or enjoy the privileges of membership.

complex language. Only if you can use language in a complex way are you suitably qualified to occupy a privileged position in society. Furthermore, to impose another barrier, prospective specialists in a field have to prove their ability under artificially secluded conditions. They have to write and pass examinations ...

TIME TO REFLECT

What do you think about the points made so far in this section? Write down your answers to the following questions to see where you stand and what you think:

✪ Do you think that the description of education given above is too cynical or one-sided?
✪ Do you think that the government of a country should have a say in what is taught at school and university? If you answer yes, to what extent should the government control education? If you answer no, who should decide on what is taught and how it is taught?
✪ Do you think that your writing in essays and examinations is an accurate indication of your intellectual ability or your social value?
✪ Are you encouraged to express your own ideas and opinions in the courses that you are studying or are you expected to reproduce the ideas and opinions that are held by your teachers?
✪ When you are asked to give an opinion, do you think that your views are respected?
✪ Do you think that universities in your country are doing enough to promote social equality and upliftment?

✪ You may have heard arguments calling for students from politically and economically disadvantaged backgrounds to be given special treatment by university authorities. For example, it has been suggested that universities should lower their entrance requirements or alter their examination standards for people who have previously been the victims of discrimination and oppression. What do you think of these views? Why?

. . . .

FOUR

MICHEL FOUCAULT: ON KNOWLEDGE AND POWER

THE GROUNDBREAKING AND controversial writings of Michel Foucault (1926–1984) have had a huge influence on our modern view of knowledge and power. Although his work covers many fields in the human sciences – philosophy, history, psychology, medicine, gender studies and literary criticism – its objective has, in Foucault's words, been 'to create a history of the different modes by which, in our culture, human beings are made subjects' ('The Subject and Power' in Herbert et al., 1982. *Michel Foucault: Beyond Structuralism and Hermeneutics*, p. 208. Brighton: Harvester Press). When Foucault speaks of human beings as 'subjects' he is using the word in two ways. First, and significantly, he is referring to the manner in which humans have become the subject of scientific study as in psychiatry, psychology and medicine. Secondly, his focus is on the processes by which human beings permit themselves (or are coerced into) to be governed. In other words, he asks: How do individuals become subject to authority in society, whether that authority is governmental, legal, religious or educational? The inseparable connection between knowledge and power is a crucial aspect of both uses of the word, 'subject'. Underlying all of Foucault's work is the argument that although knowledge and power are not the same thing, each incites the production of the other. Knowledge gives rise to power. Power engenders knowledge. It is impossible for

● IF KNOWLEDGE cannot be separated from power or power from knowledge, how can we tell the difference between good knowledge and bad knowledge?

power to be exercised without knowledge. At the same time, power promotes particular forms of knowledge.

The logic of this argument is not easy to grasp, but it has had an enormous impact on our concept of 'truth'. If we accept Foucault's position, we can no longer assume that objective knowledge is possible; that 'truth' and the 'true untainted nature of humanity' exist somewhere beyond politics and society. What is regarded as 'true' is always determined by power relations. Furthermore, because everyone is caught up in a net of power relations, there is no position from which we can observe the world that is free from particular expressions of power. All statements are determined by power and result in power.

To make these ideas clearer we need to look more closely at Foucault's work. As you read through the discussion, think back to the points that have been made so far in this chapter and in this book. Do Foucault's arguments add to your understanding of the claims that we have made? Or, are there differences between these arguments and the ideas expressed in *Love, power and meaning*?

Madness and Medicine

● HOW DO WE
know that mad
people are mad?

A central theme in all Foucault's work and the focus of his early studies is the categorization of people into 'normal' and 'abnormal'. We naturally tend to define abnormality as a condition or state that differs significantly from what is considered normal. Furthermore, we tend to assume that the difference between abnormality and normality is easy to define and that it remains the same over time. Foucault challenges this. He shows that definitions of abnormality vary greatly over time. He also demonstrates that the changing sense of abnormality is accompanied by a growing scientific interest in the category and that this interest is closely linked to power in society. We only know what is normal by defining what is abnormal; and, once these categories are defined it is always the 'normal' person who has power over the 'abnormal'.

In *Madness and Civilization: A History of Insanity in the Age of Reason* (1965), Foucault argues that the category of madness operates as a mechanism for excluding certain types of people from society, especially by confining them in prisons or institutions. He claims that during the Middle Ages people whose behaviour deviated from the norm were tolerated, while people with contagious diseases (such as leprosy) were incarcerated or banished. By the middle of the fourteenth century this began to change. With the Renaissance, madmen and vagabonds came to represent a danger to society. But the social response to this danger was ambiguous. On the one hand, mad individuals were rejected and ritually excluded. On the other hand, they were seen as an essential part of society because they understood the mysteries of human existence. Significantly, in Renaissance literature the fool and the madman often speak truths that cannot be spoken by anyone else. By the seventeenth century, though, deviation was no longer tolerated. Even in art and literature it was represented as a disruptive menace to social order. Anyone who was regarded as a troublesome element in society was confined in a hospital, prison or institution. Moreover, what constituted troublesome behaviour extended far beyond criminality and madness. The unemployed, the sick and the aged were segregated from society along with criminals and madmen.

These changes in the treatment of madness were, according to Foucault, not because of chance or the gradual evolution of society. Instead, they are directly linked to knowledge and power. In particular, the great changes of the seventeenth century can be traced directly to the economic crisis that affected most of Western Europe at that time. The detention of troublesome elements was a means of protecting social order against the

FURTHER READING

For a light and readable overview of Foucault's thinking try:

Lydia Alix Fillingham, 1993. *Foucault for Beginners*. New York: Writers and Readers Publishing.

For a more detailed (and more difficult) account you could consult:

Philip Barker, 1998. *Michel Foucault: An Introduction*. Edinburgh: Edinburgh Univerity Press.

Lois McNay, 1994. *Foucault: A Critical Introduction*. Cambridge: Blackwell.

Or use the name 'Foucault' in a search of your library catalogue or the Internet.

● FOUCAULT SHOWS that at a certain point in the seventeenth century one in every hundred citizens of Paris found themselves in a place of confinement.

threat of uprisings and civil disturbances. At the same time, it created a useful pool of cheap labour. More importantly, confinement formed a new understanding of the relation between work and morality. The ability to work productively became a way of defining what was socially normal. Laziness was seen as a form of rebellion and the proper way to treat the idle was to put them to work. Similarly, madness was seen as a failure to integrate into what was normal in society. By lumping madness together with all sorts of socially and economically undesirable traits, the mad lost their former status as visionaries or speakers of essential moral truths and became simply part of the social 'other' – the alien group against which the socially 'normal' people defined their normality.

There is a second motive for segregating and confining mad people – reason. As we noted when we were discussing Locke and Rousseau earlier in this chapter, reason (or the capacity for rational thought) was regarded by seventeenth-century philosophers as the characteristic that defined human nature and set it apart from the rest of the animal kingdom. Madness, measured against the social norm, is the opposite or antithesis of rationality. It operates outside the boundaries of rationality; it is inherently illogical. If the essence of humanity is rationality, then madness contradicts the assumption that humans are rational creatures. By excluding all forms of 'unreasonable' thought, by isolating and 'othering' those individuals who behave irrationally, society is able to pretend that rationality is its organizing principle. Quite literally, mad people are removed from society and kept hidden. With the presence of madness obscured, those who are left behind can believe that they and their society are intrinsically rational.

Foucault's analysis goes on to show how the treatment of madness placed the moral and ethical responsibility for their condition on mad people themselves. The origins of their dysfunction were to be found not in society but within the bodies and minds of the insane. Simply put, mad people were blamed for their madness. Initially, treatment was focused on the body.

Later, especially with the emergence of psychotherapy, the mind became the object of treatment. This shift in responsibility is extremely important for the relationship between knowledge and power, primarily because it places people in a position where they are expected to monitor and control themselves through constant self-examination. The individual is required to be self-regulatory. By emphasizing personal responsibility some of the burden for social control is removed from the State. Self-awareness and self-control are also essential components of the cure. The patient has to demonstrate that he or she has internalized the norms and ethics of society, is no longer a disruptive element, and can resume a place in society without posing any threat to social order. And the professional doctor, wielding scientific knowledge, is instrumental in the diagnosis, treatment and pronouncement of a cure. In subsequent studies, including *The Birth of the Clinic: An Archaeology of Medical Perception* (1963) and *The Order of Things* (1966), Foucault shows how intense observation, new systems of classification and the increasing importance of the human sciences are closely linked to maintaining power relationships in society.

Foucault shows how mad people were blamed for their madness.

● DO YOU agree with Foucault's assessment of psychotherapy?

We have used Foucault's account of madness to explain broadly how knowledge (especially knowledge that determines what is normal and abnormal) can be used to exercise power. We have also shown how power, represented by social norms and implemented by social apparatuses, can work to privilege certain types of knowledge. The downfall of authoritarian forms of government gave rise to the need for more subtle and sophisticated means of social control. These forms of social control, in turn, led to an increasing emphasis on studying and knowing the individual. The transition from a society that operates by brute force to a society that functions in less obvious ways is the subject of Foucault's later works, to which we are now going to turn.

Punishment and Sexual Pleasure

Foucault's analysis of knowledge and power is extended in two of his most influential studies, *Discipline and Punish* (1975) and *The History of Sexuality, Volume I: An Introduction* (1976). For the most part, I'll concentrate on the first work, saving Foucault's discussion of contemporary sexuality for my concluding comments.

The opening paragraphs of *Discipline and Punish* (first published in French as *Surveiller et punir*) are probably the most astonishing of any book on modern social philosophy:

> On 2 March 1757 Damiens the regicide was condemned 'to make the *amende honorable* before the main door of the Church of Paris', where he was to be 'taken and conveyed in a cart, wearing nothing but a shirt, holding a torch of burning wax weighing two pounds'; then, 'in the said cart, to the Place de Grève, where, on a scaffold that will be erected there, the flesh will be torn from his breasts, arms, thighs and calves with red-hot pincers, his right hand, holding the knife with which he committed the said parricide, burnt with sulphur, and, on those places where the flesh will be torn away, poured molten lead, boiling oil, burning resin, wax and sulphur melted together and then his body drawn and quartered by four horses and his limbs and body consumed by fire, reduced to ashes and his ashes thrown to the winds' (*Pièces originales...*, pp. 372–374)

> 'Finally, he was quartered,' recounts the Gazette d' Amsterdam of 1 April 1757. 'This last operation was very long, because the horses used were not accustomed to drawing; consequently, instead of four, six were needed; and when that did not suffice, they were forced, in order to cut off the wretch's thighs, to sever the sinews and hack at the joints ...

'It is said that, though he was always a great swearer, no blasphemy escaped his lips; but the excessive pain made him utter horrible cries, and he often repeated: "My God, have pity on me! Jesus, help me!" The spectators were all edified by the solicitude of the parish priest of St Paul's who despite his great age did not spare himself in offering consolation to the patient.'

(FROM: Michel Foucault, 1991. *Discipline and Punish: The Birth of the Prison*, p. 3. Harmondsworth: Penguin.)

In contrast to Damiens's punishment is Léon Faucher's routine for prisoners in Paris in 1837, a mere eighty years later:

'Art. 17. The prisoners' day will begin at six in the morning in winter and at five in summer. They will work for nine hours a day throughout the year. Two hours a day will be devoted to instruction. Work and the day will end at nine o'clock in winter and at eight in summer.

'Art. 18. *Rising*. At the first drum-roll, prisoners must rise and dress in silence, as the supervisor opens the cell doors. At the second drum-roll, they must be dressed and make their beds. At the third, they must line up and proceed to the chapel for morning prayer. There is a five-minute interval for each drum-roll. (1991: 6)

This continues, until each hour of the day is accounted for in a timetable of activity. Although Foucault acknowledges that the two punishments do not apply to the same crime, he is interested in the change from punishment centred on the body to punishment that controls every aspect of life. How and why did this happen? For Foucault, this does not mean that more humane notions of punishment were adopted. Rather, it represents a refinement of social control – a more efficient and effective way of exercising power. According to Foucault, punishment must be seen as serving a social function by acting both repressively and by encouraging positive effects. At the same time, it is a political tactic because it is a particular way of exercising power.

Before the end of the seventeenth century, pain was an

important part of the judicial process. Torture was used to establish guilt, and pain was integral to the punishment of the offender. By acting directly and publicly on the body of the accused or guilty party, the judicial process left visible traces of the results of crime. Quite literally, the body was marked or scarred. In addition, public punishments – especially executions – demonstrated the power of the king or sovereign. Crime was seen as an attack on the sovereign because it represented a challenge to the sovereign's ability to maintain order. Punishment, torture and public execution were manifestations of the sovereign's power over life and death. Through ritualized punishment, such as the one meted out to Damiens, revenge could be taken and order restored.

But the public nature of punishment also had its risks. Public displays of institutional violence could lead to the crowd becoming unruly and rioting, especially if they felt that punishment was excessive or unjust. On several occasions the crowd had disrupted the public display of power by preventing the execution of the condemned person or by mocking and jeering. Instead of asserting the sovereign's authority the event had the potential to undermine it.

For Foucault, the threat of disorder led to penal reform. Ideally, punishment should deter criminal activity without seeming arbitrary or disproportionate to the crime. The sheer violence of the public spectacle had to be replaced by a graded system of punishment, disassociated from the king's authority but nevertheless satisfying the social requirement that criminals should be made to pay for their crimes. The modern prison or penitentiary best serves these purposes. The prison system becomes a new way of exerting power.

The form imprisonment takes is also intricately linked to knowledge. Penal reform coincides with the rise of Enlightenment humanism, which in turn created new ways of studying and controlling human behaviour. The principles of these new human sciences were ideally suited to a changing view of disciplinary power, because they are based on classifying, identifying and arranging human beings and human bodies.

Likewise, disciplinary power in prisons and other social institutions, such as schools, factories and hospitals, relies on the classification and control of bodies and behaviour. Outside of the prison, the military is the closest approximation to the way in which disciplinary power operates. In the military, and in disciplinary organizations, everyone is allocated a specific place or position. The activities of people are minutely controlled, often by timetables. These activities involve repetitive and standardized exercises, which must be carried out by everyone. There is an elaborate chain of command or hierarchy of power. Each level keeps watch over the level below it. Lastly, there is strong normative or internal control. The individual is constantly monitored by, and against, the backdrop of the group. In this way, any deviation from the norm can be quickly identified and punished, while conformity can be rewarded. Therefore, tactics of disciplinary power discourage and correct undesirable behaviour. They allow effective control with a minimal risk of social disruption. Moreover, because the strategies of disciplinary power are applicable across a wide range of cultural institutions, its effects can be total. From the eighteenth century, correct training, close monitoring and careful examination (whether in schools, prisons or mental institutions) become mechanisms for fixing the positions of individuals and exercising power over them.

Foucault draws on Jeremy Bentham's (1748–1832) architectural design for the Panopticon as a particularly intense example of how the new disciplinary regime works. A Panopticon is a circular, multi-storeyed building that is built around a central tower. Here is Foucault's description:

This tower is pierced with wide windows that open onto the inner side of the ring; the peripheric building is divided into cells, each of which extends the whole width of the building; they have two windows, one on the inside, corresponding to the windows of the tower; the other, on the outside, allows light to cross the cell from one end to the other. All that is needed, then, is to place a supervisor in the central tower and to shut up in each cell a madman, a

peripheric building: the building around the central tower

patient, a condemned man, a worker or a schoolboy. By the effect of backlighting, one can observe from the tower, standing out precisely against the light, the small captive shadows in the cells of the periphery. They are like so many cages, so many small theatres, in which each actor is alone, perfectly individualized and constantly visible. The panoptic mechanism arranges spatial unities that make it possible to see constantly and to recognize immediately. In short, it reverses the principle of the dungeon or rather of its three functions – to enclose, to deprive of light and to hide – it preserves only the first and eliminates the other two. Full lighting and the eye of the supervisor capture better than darkness, which ultimately protected. Visibility is a trap. (1991: 200)

● **TRY** to imagine that you are imprisoned in a Panopticon. How would you feel? What would you fear? Why would 'visibility' be 'a trap'?.

'Visibility is a trap' because the inmate can be observed at all times without being able to see the observer. It doesn't matter whether observation is continuous or not. It also doesn't matter who is doing the observing. The inmate must act as though the observation is taking place at all times, or suffer the consequences of being caught breaking the rules. Thus, the consciousness of permanent visibility allows power to function automatically. The mechanism (in this case the Panopticon) that creates and sustains power is independent of the person who exercises it. In Foucault's words, 'the inmates [are] caught up in a power situation of which they are themselves the bearers' (1991: 201). Furthermore, although the inmates are obviously in a position of less power than the observer in the tower, the central observers are also subject to control. They too have a fixed position and a regulated set of functions. Thus, even as they control others, they are subject to control.

The Panopticon is more than just a very clever prison. It also works as a metaphor for the operation of disciplinary power in Western culture in general: it is a model of the way that power relations function in terms of everyday life. Think of the use of surveillance cameras in shopping malls and city centres. The monitoring of the camera is continuous, disciplinary and anonymous. Anyone can operate it. Even if no one is looking at the screen it exerts control because the people who are being observed have no way of telling whether they are being seen or not. Consequently, they become responsible for controlling their own behaviour. The anonymous individual tucked away in some central observation point exercises power over the crowds of shoppers. However, because the individual has a defined position and a set of duties, he/she is an easy target for control. Power may be distributed unevenly through society, but its effects are inescapable. Finally, the condition of constant visibility and subtle control is accepted as normal, even socially desirable. No one objects to the process of surveillance because they see it as being in the interests of their personal security.

The architectural model of the Panopticon also applies to knowledge, especially in terms of its attempts to transform individuals. For a moment, let's imagine that the observer in the tower is a scientist operating in one of the human sciences: say psychology, sociology or anthropology. The building around the tower is society and each of the cells represents an individual. From a central observation point, the scientist can easily observe the field of study. He or she can order the individuals in their cells, classify them, rearrange them, require them to perform particular tasks and subject them to experiments. Anything out of the ordinary, any strange behaviour, any departure from the norm is instantly recognizable. If individual *X* in Cell *B* shows the slightest sign of deviation appropriate steps can be taken to correct his or her aberrant behaviour. Paradoxically, the human sciences make it possible to recognize individuality in way that was not possible

WAS YOUR SCHOOL A PANOPTICON?

- How was the school of pupils divided up?
- How were classes arranged?
- What did your classroom look like? Did it have windows of equal size on both sides?
- Did you have a fixed place where you had to sit?
- Who had the most power in the school?
- Who had the least power?
- Did some pupils have more power than others?
- How was your school day arranged?
- How was your school week and year arranged?
- Who decided what work was done and when?
- How was that work assessed?
- What happened if you did badly at school or broke the rules?
- What happened if you did well and kept to the rules?
- Do you think the principles of the Pantopticon were used in organizing the structure of your school?

before. Each category of humanity and human behaviour can be named, placed, described and treated. However, at the same time, intense observation and clear categorization allows strong normative or internal control. Knowledge allows the individual to emerge from the undifferentiated mass of humanity, and, as it does so, it allows power to be exerted in a manner that ensures conformity to the norm.

Interestingly, although it influenced the design of prisons, Bentham's Panopticon was never built. Its principles, though, had a profound effect on the administration and control of society. Because the Panopticon is an ideal way of connecting bodies, space, power and knowledge, it has informed the shape and functioning of many modern social institutions. Given its efficiency and effectiveness, Foucault asks: 'Is it surprising that prisons resemble factories, schools, barracks, hospitals, which all resemble prisons?' (1991: 228)

Foucault goes on to show that the emergence of new systems of power coincides with the appearance and growth of the categories of anomalies – 'abnormal' behaviour or individuals, such as the 'delinquent', the 'pervert', the 'homosexual' – that the technologies of power were supposedly designed to destroy. Normalization (i.e., the process of causing to conform) works by labelling certain things 'normal' and others 'abnormal'. Here, we need only consider how notions of hysteria in women or homosexuality (predominantly in men) increasingly emerged as aberrant or 'abnormal' categories for identification and treatment during the late-nineteenth century to recognize the productive nature of the new science of psychotherapy. It is not true that women with emotional problems and male homosexuals did not exist before they were scientifically studied, categorized and treated. They have always been a part of society. However, the label of 'abnormality' justified a particular response that is directly related to social control. If we identify women as naturally unstable, then it is far easier to condone a political or social system that strips women of their power. Similarly, if the family is an important means of keeping social cohesion, then labelling those who transgress the ideas of normal sexual relations is

● IF the tactics of disciplinary power are so pervasive in our society, how is resistance to control possible?

POWER

extremely useful. Labelling criminalizes what is essentially private behaviour. It is a powerful mechanism for ensuring that this behaviour is suppressed or concealed. Moreover, because these abnormalities are identified 'scientifically' they can be controlled more effectively. What is essentially a political problem is recast in the 'neutral' language of science. It is thus possible to establish patterns of political control without appearing to do so. Instead, the actions of the dominant group on those who break the norm are labelled 'reform' or 'therapy'. But they serve a more sinister purpose: to preserve the political and economic status quo.

This brings us to one of the most interesting insights of Foucault's analysis. Creating categories in order to exert control is never simply repressive. It does not have to be enforced from the top down. Instead, power is exerted in all directions. People police themselves and they police others, who also form part of a policing network in which surveillance and control are everywhere. But for every act of power, there is a possible act of resistance; and the possibility of resistance also, in some respects, depends on the link between knowledge and power. Take homosexuality, for example. Before it was systematically analysed and categorized, it was difficult to control homosexual behaviour. Once it had been labelled and turned into a pathology (or illness) it could be treated as a social evil by the criminal justice system, by the medical profession and by religious institutions. But the process of labelling had multiple effects. It allowed political action to be taken against homosexuals, but it also allowed homosexuals to identify themselves as part of a coherent group. Labelling gave what was then merely a particular behaviour a social identity. In time, this identity, linked to social repression, caused homosexual people to redefine themselves in a way that resisted negative labelling. Clinical categories of abnormality were overtaken by a new sense of self and a new vocabulary – the word 'homosexual' gave way to the term 'gay' or to the adoption of the negative label 'queer'. At times, passive compliance with social norms was replaced by acts of physical resistance. Thus, power is always countered by the possibility of resistance.

HOMOSEXUALITY AND RESISTANCE

The Stonewall Riot in New York City in 1969 was the first public protest by gay people against police and social harassment. Over time, resistance took the form of high-profile campaigns calling for equality by organizations such as the National Gay and Lesbian Task Force in America and the National Coalition for Gay and Lesbian Equality in South Africa. Significantly, the assertion of gay and lesbian identity followed closely on the publication of new studies into human sexuality: *Sexual Behaviour in the Human Male* (1948) and *Sexual Behaviour in the Human Female* (1952), by American biologist Alfred Charles Kinsey. Although these works contained inflated estimates of the size of the homosexual population and the incidence of behaviour, they provided a more realistic picture of homosexuality and helped demystify it. Unlike earlier studies, which focused on individuals who had sought medical or psychological help, the Kinsey reports described homosexuals outside of clinical settings. His study showed that gay people were to be found in all walks of life, growing up in all kinds of families, and practising many different religions. As a result of the ensuing scientific discussion, the American Psychiatric Association in 1973 eliminated homosexuality from its list of mental disorders and, in 1980, dropped it from its *Diagnostic and Statistical Manual*.

Foucault argues that prisons do not usually succeed in reforming prisoners. During their sentences they become involved with other criminals and leave prison only to commit further crimes.

✪ Do you see recidivism (committing crime again) as an admission of the failure of the modern prison system?

✪ Or, do you agree that the State expects recidivism and that it is built into the system? Do you think that some people will always offend (they are criminal by nature) and the best place for them is in prison where they are separated from society and can do relatively little damage?

✪ In many Islamic countries, criminals can have a limb cut off in public as a penalty for their crimes. Public executions also take place. Do you think that this is a more effective method of discouraging criminal behaviour and punishing offenders than prison?

✪ Do you think the death penalty should be instituted for serious crimes? Why or why not?

At the close of the twentieth century, gay people in many parts of the world find themselves protected by constitutions and legal systems that previously oppressed them. In other contexts, they are still labelled 'abnormal'. Social acceptance comes at a price, though. By moving from the periphery of abnormality to the social centre, gay people have found themselves under pressure to adopt the social and moral codes of the wider society. Thus whether they are excluded as a manifestation of abnormality or whether they are included as an integral component of society, as a group they remain under careful control. Such is the nature of knowledge and power.

What is Power? Foucault's Answer

One might expect that a writer whose work is about the relationship between knowledge and power would offer his readers a grand and complete theory of power. Foucault never does. For him, power is not a thing, an institution, an aptitude or an object. It describes relations of force in particular social bodies and is always linked to specific contexts, conditions and histories. One of the reasons that we have dipped into the details of his studies of madness and imprisonment is that various power relations are inseparable from the social bodies that give them meaning. However, in *The History of Sexuality* Foucault does offer a series of ideas that are useful in summarizing his sense of power and that may be useful for consolidating our concept of power as we approach the end of this chapter.

- Power relations are non-egalitarian and mobile. The way that power functions in society is never equal, nor is it fixed in time or space. Some people have more power than others, but no person is free of the effects of power.
- Power is not restricted to political institutions. Relations of power are interwoven with other kinds of relations: work, family, sexuality. Every specific individual occupies various positions in networks of power and so stands in multiple positions. For instance, the same individual may be a brother, father, friend, teacher and employee. Each position carries with it its own power relations. Power therefore cannot be a permanent one-way exchange. It does not flow uniformly from the more powerful to the less powerful. Rather it circulates among bodies.
- Power does not take the sole form of prohibition and punishment but is multiple in form. It exerts control through repression and also through reward. It offers both the possibility of conforming and resistance.
- Power relations exist because they are capable of being used in a wide range of strategies. Significantly, power relations

are never just a matter of chance. They are intentional and subjective. At a local level there is often a high degree of conscious decision-making and political manoeuvring. Nevertheless, the overall effect of power escapes the intentions of any individual actor. The sum total of multiple intentions is greater than the individual parts.

✪ There are no relations of power without possible resistances.

. . . .

FIVE

LITERARY POLITICS

YOU MAY BE wondering what all this has to do with literature or with English studies. Why are we interested in power in the first place? And, why have we dealt with it in such detail and from so many angles? How will all these ideas increase our capacity to read and to write? These are questions that I turn to in the final chapter of *Love, power and meaning*, when I look at politics and literature. Before you get there, though, I'd like you to read part of an essay by Ngũgĩ wa Thiong'o – a highly-respected Kenyan novelist and literary critic. As you read, think about the arguments Ngũgĩ makes. Do you agree with his viewpoint? What are the implications for literature and literary criticism of accepting his sense of the connection between writer and society?

● DOES NGŪGĪ'S use of 'men' and 'man' exclude women's literary achievements?

Literature results from conscious acts of men in society. At the level of the individual artist, the very act of writing implies a social relationship: one is writing about somebody for somebody. At the collective level, literature, as a product of men's intellectual and imaginative activity embodies, in words and images, the tensions, conflicts, contradictions at the heart of a community's being and process of becoming. It is a reflection on the aesthetic and imaginative planes, of a community's wrestling with its total environment to produce the basic means of life, food, clothing, shelter, and in the process creating and recreating itself in history.

It is important to make that simple and obvious point at the beginning because the general tendency is to see literature as something belonging to a surreal world, or to a metaphysical ethereal plane, something that has nothing to do with man's more mundane, prosaic realm of attempting to clothe, shelter and feed himself.

At the same time literature is more than just a mechanistic reflection of social reality. As part of man's artistic activities, it is in itself part of man's self-realization as a result of his wrestling with nature; it is, if you like, itself a symbol of man's creativity, of man's historical process of being and becoming. It is also an enjoyable end-product of man's artistic labour. But more important, it does shape our attitudes to life, to the daily struggle with nature, the daily struggles within a community, and the daily struggle within our individual souls and selves.

It follows then that because of its social character, literature as a creative process and also as an end is conditioned by historical social forces and pressures: it cannot elect to stand above or to transcend economics, politics, class, race, or what Achebe calls 'the burning issues of the day' because those very burning issues with which it deals take place within an economic, political, class and race context. Again because of its social involvement, because of its thoroughly social character, literature is partisan: literature takes sides, and more so in a class society.

A writer after all comes from a particular class and race and nation. He himself is a product of an actual social process – eating, drinking, learning, loving, hating – and he has developed a class attitude to all those activities, themselves class conditioned. A writer is trying to persuade us, to make us view not only a certain kind of reality, but also from a certain angle of vision often, though perhaps unconsciously, on behalf of a certain class, race, or nation:

> All art aims to evoke; to awaken in the observer, listener or reader emotions and impulses to action or opposition. But the evocation of man's active will requires more

than either mere expression of feelings, striking mimesis of reality, or pleasing construction of word, tone or line: it presupposes forces beyond those of feeling and form which assert themselves simultaneously and in harmony with emotional forces, fundamentally different from them. The artist unfolds these forces whether in the service either of a ruler, despot or monarch – or of a particular community, rank in society or financial class; of a state or church, of an association or party; or as a representative or spokesman of a form of government, a system of conventions and norms: in short, of a more or less rigidly controlled and comprehensive organization. (Arnold Hauser, 'Propaganda, Ideology and Art', in *Aspects of History and Class Consciousness*. Ed. Istvan Meszaros, p. 128.)

Seen in this light, the product of a writer's pen both reflects reality and also attempts to persuade us to take a certain attitude to that reality. The persuasion can be a direct appeal on behalf of a writer's open doctrine or it can be an indirect appeal through 'influencing the imagination, feelings and actions of the recipient' (Arnold Hauser, p. 129) in a certain way toward certain goals and a set of values, consciously or unconsciously held by him. A nation's literature which is a sum total of the products of many individuals in that society is then not only a reflection of that people's collective reality, collective experience, but also embodies that community's way of looking at the world and its place in the making of that world. It is partisan on the collective level, because the literature is trying to make us see how that community, class, race, group has defined itself historically and how it defines the world in relationship to itself.

(FROM: Ngũgĩ wa Thiong'o, 1981. 'Literature and Society' in *Writers in Politics*, pp. 5–7. London: Heinemann.)

6 The politics of reading [and] writing

THE FINAL CHAPTER of *Love, power and meaning* differs from previous chapters. First, it focuses mainly on a single literary work – a one-act play by Zakes Mda. Second, apart from offering strategies for 'reading' literature, it intends in a very explicit and self-conscious way to draw attention to the political implications of the web of practices that make up the literary scene. In other words, in this chapter I have set out to look at the politics of reading writing – specifically reading literature – and at the politics involved in writing both literary texts and texts about literature. Finally, in discussing the political dimensions of literature and literary criticism, I haven't pretended that I am able to leave behind my personal ideas about literary value or acceptable politics. Instead, I have consciously tried to reflect on the strengths and weaknesses of my own assumptions and positions.

Foucault claims that his analysis of power has several key lessons for literary and cultural critics. Among these lessons, three are of particular importance to this chapter. First, Foucault argues that we should analyse those things that seem closest to us; we should critically examine things that appear self-evident or unquestionable. Second, he claims that we should explicitly undertake our analysis from our own perspective and avoid speaking on behalf of others who are silenced by our speech. Finally, we should direct our attention to those forms of domination that appear to be globalized. By this, Foucault means that we should focus on struggles against oppression, domination and exploitation, whether these are based on gender, race, sexual orientation, religion, class or economic status.

In the light of Foucault's arguments, a claim on my part for critical objectivity and political neutrality would be out of place. Instead, I have tried to explore what I am really saying about myself when I make particular claims for politics, literature, the politics of literature, and political literature. More importantly, by using myself as an example I want to encourage you, the reader, to examine your own attitudes and responses with greater self-consciousness. This implies that you will have to participate actively in making this chapter meaningful. Although I will provide the main framework of materials and resources, the emphasis is on your own reading, research, thinking and writing.

Let's get to work by looking at the play.

Dark Voices Ring

for Dini

Dark Voices Ring was first presented at the People's Space Theatre, Cape Town, on 9 October 1979 with the following cast of characters:

WOMAN : **Nomhle Nkonyeni**
MAN : **Nkos'omzi Ngcukana**
OLD MAN : **Sam Phillips**

Directed by Nomhle Nkonyeni and Rob Amato

Inside a hut. An old weather-beaten man sits on the floor. His whole body is covered in a tattered blanket. He doesn't seem to have any limbs, although we can't be sure about this because of the blanket. Only his skull-like head is showing. He is gazing fixedly into empty space – and he is going to remain in this position for most of the play. Beside him sits a woman. She seems to be mourning since she is wearing a black doek and a black dress. She also has a black shawl over her shoulders. She is mumbling something which is inaudible to the audience. She seems to be uneasy and she fidgets. Now and then she stares at the empty space that the old man is gazing at, but seeing nothing she hopelessly goes back to her fidgeting and mumbling. A man appears from behind, looks at them, shakes his head and smiles broadly.

MAN *(calling out in a piercing but friendly voice)*: Ho-oo! Ho-o-ooo-! Ho-o!
> WOMAN *starts, but* OLD MAN *does not move. He continues to gaze into empty space.*

WOMAN *(standing up)*: Why did you do that?

MAN: Do what?

WOMAN: Ag, forget it. *(There is an awkward pause)* Where did you leave Nontobeko?
> MAN *pretends not to hear.*
> I tell you every day that when you come to visit me you should not leave my child at home. Times are dangerous. You never know what maniac might attack her ... left alone in that house of yours. And her father *(pointing at OLD MAN)* wants to see her too.

MAN: The old man says he wants to see Nontobeko?

WOMAN: Yes. It's been a long, long time since we saw her.

MAN: You mean he spoke to you and said he wants to see her?

WOMAN: He is my husband. He speaks with me, and I know that he wants to see his daughter as much as I want to see her.

MAN: I haven't heard him speak since he came from the potato farms.

WOMAN: You are not his wife.

MAN: Come off it, mama. You know that the old man doesn't know how to speak. Speaks no evil, sees no evil, hears no evil.

WOMAN: If you can't hear him I can. Not the words. His words are lost. They got lost many years ago in the farms where his manhood was also buried together with the manure that makes the potatoes big and rich. No. Not his words. There are so many things he could say – the pain and the anguish of a lost home ... lost hope – but he can't say them with words. The old man speaks to me. Even now I can hear his

cry like the cry of a baby wrapped in newspapers and abandoned in a lavatory pit.

MAN (*incredulously*): I don't want to say you are lying ma, but I
don't believe you can hear his cries. Imagine them ... yes.
Dream of them perhaps. But hear them ... impossible.

WOMAN: I don't live on dreams, child. If I did I would have gone
berserk with worry long ago. (*Confidentially*) With neighbours like ours one cannot be too careful with dreams.

MAN: I am talking of decent dreams ... not the kind of dream you
fear.

WOMAN: I have experienced how decent dreams will turn dirty at
the bat of an eye. A sweet, handsome man may turn into a
raving maniac. No, child, there are no decent dreams.

MAN: You can dream of the good times ... when Nontobeko was
born. It must have been a happy event for you and the old
man.

WOMAN: In truth it was. Our first child, Nontobeko. The pains of
birth started in the morning when I was weeding our vegetable patch. I was a very strong woman in those days, my
child. Even when Nontobeko was kicking in my belly I used
to work from sunrise till dusk.

MAN: You have always been strong, ma. My mother never forgets
how you used to gather the biggest bundle of wood when
you were girls. Much older women couldn't lift your bundle to their heads.

WOMAN: She was rather violent. Kicking for all the world to see.

MAN: My mother had a fiery temper.

WOMAN: I am talking about Nontobeko.

MAN: Oh, yes. The coming of Nontobeko into the world.

WOMAN: The coming of Nontobeko to Baas Jan van Wyk's potato farm. She kicked her way into the world, that one.

MAN: A beautiful coming. It would make a decent dream.

WOMAN: Her father always wanted a boy as a first child.
Somebody who would take over from him in his old age.

MAN: Someone who would inherit the labour of the farm. The
old man must have been very ambitious in his day.

WOMAN: Many fathers like their sons to follow in their footsteps.

The old man was not different. He was a *baas-boy* on Jan van Wyk's farm. Van Wyk was giving him a bag of mealie-meal every month. We had our own vegetable patch, our own hut, and our own milk cow. Besides, all the other farm labourers were under the old man's command. Kaptein, they called him. When he called them they all used to answer: *Yebo nkosi ... ewe nkosi ... morena.* The master of the farm called him 'my faithful *induna*'. The old man wanted his son to inherit his position.

baas-boy: a racist term for overseer

kaptein: captain
Yebo nkosi ... ewe nkosi ... morena: Yes sir ... yes sir ... my lord
induna: headman

MAN: To inherit the power and the glory and the kingdom of Baas Jan van Wyk's farm.

WOMAN: It was before the master brought prisoners to dig out the potatoes with their bare hands. All the digging was done by farm labourers. The old man was on the farm inspecting the work they did.

MAN: Driving them to work even harder.

WOMAN: And I was weeding the vegetable patch. Then I felt the pains. At that time I did not know they were the pains of birth – I had never given birth to anyone before. I tried to ignore the pain, and continued to weed my vegetables.

MAN: Beautiful vegetables that drove the pellagra away from the household of the Kaptein.

pellagra: a disease caused by the absence of nicotinic acid

WOMAN: After some time I went home to prepare the midday meal which I had to take to him in the fields.

MAN: To give more strength to the hand that was wielding the whip.

WOMAN: The pain was now unbearable. Nontobeko was kicking hard in my bowels.

MAN: And I suppose at the same time she was screaming: 'Let me out! Let me out into the farm of Baas van Wyk, into the kingdom of my father!' She must have been very eager to take over the sacred duty of running the farm.

WOMAN: It was as if a thousand razor blades were cutting my insides. All the same I went on, all the way to the fields. I found the old man in a bad temper. I knew at once that the farm labourers had been giving him a lot of trouble. I did not tell him about the pain. He never liked family problems to interfere with his work.

MAN (*dials on a imaginary phone*): Hello. I am Kaptein's wife. Yes, *baasie*, the wife of the faithful *induna*. May I make an appointment with him for five o'clock? Tomorrow afternoon, yes ... Well, *baasie*, I would like to discuss with him family affairs ... Pains in my stomach ... Oh, yes. I live with him all right ... Yes, in the beautiful hut you built us ... but you know how these things are. He is always busy with the problems of your farm. Thank you, *baasie*. Tomorrow five o'clock, then.

WOMAN: After he had eaten I took the dishes and the billycan and slowly walked back home. I had not walked a long distance when the pain came again, tearing me to pieces. I was now giddy and my knees could carry me no longer. I fell down, and what happened after that is a matter of they-say.

MAN: Behold! Nontobeko is knocking at the door of the world, but no one is ready to open it. So now she is gate-crashing.

WOMAN: They say I was found lying there by passing women who were also taking food to the farm labourers. Fearing that something very serious had happened to me they ran with all their strength to call the missis. By the time they came back with her, Nontobeko had come half-way into the world.

MAN: She must have been a very impatient girl.

WOMAN: They took me – not to my hut, mind you – to the *huis* of Jan van Wyk which is where the other half of Nontobeko was born. The missis herself was the midwife.

MAN (*to the audience*): Nontobeko was born in no simple mud hut. She was born in the *huis* of the master of the farm.

WOMAN: She has always lived with that prestige. The only child on the farm, from the beginning of time, to have been born in the *huis* of the master. She was the envy of all the neighbours. Of course as the wife of the *induna* my child could not be an ordinary child.

MAN: An extraordinary beautiful dream the story of the birth of Nontobeko would be. It would make you forget the indecent dreams you fear so much.

missis: a mistress or woman of the house; wife of the boss

huis: house

POWER

WOMAN: It is beautiful insofar as it is not a dream. Once it becomes a dream the ugliness creeps in. In a dream instead of having the missis coming to take me to the *huis* I would have been surrounded by a gang of ogling farm labourers ... watching Nontobeko's struggle to break free from the dark prison of the womb ... laughing their lungs out ... whilst I sleep there ... unconscious.

ogling: staring indecently

MAN: A beautiful dream still. What is better than having the daughter of the Kaptein welcomed with cheers from a hundred of her father's subjects.

WOMAN: With jeers, my child. And in a dream you would have been there. My son-in-law watching the birth of his wife.

MAN: Since when do I feature in your dreams?

WOMAN: I feature in yours.

MAN: But mine are not ordinary dreams. Up in the northern bush people are translating them into reality – with their guns. They do it in the cities too. My dreams are dreams of freedom, ma; and you feature in them because I want you to be free.

WOMAN: I am always afraid when I feature in your dreams.

MAN: Every black man features in my dreams.

WOMAN: *Uyayibona lonto?* Just what have I been saying. Not only do you shamelessly display me in your dreams, but the neighbours have to be there too.

Uyayibona lonto?: Do you see that?

MAN: The neighbours have to be free too.

WOMAN: I wouldn't put it past you to make me appear naked in your dreams.

MAN: Unlike you, my dreams are clean, ma.

WOMAN: I am your mother-in-law, you know. I deserve a little respect from you.

MAN: I give you what is due to you.

WOMAN: You bundle me with the neighbours in your dreams. They see my nakedness. That is why I always hear them chuckle when they pass here on their way to the well.

MAN: Your life-style causes the chuckles. The old lady in never-ending mourning sitting all day long in her hut. Look ma, it's bright outside! Why don't you go out – take a walk through the village?

WOMAN: They hate me, child. Don't you realize that?

MAN: You don't give them a chance to like you. When they try to help, you say they are plotting to put *muti* in your house. So they decide to stay away from you.

muti: traditional medicine, sometimes associated with bad luck or misfortune

WOMAN: I dread going out – meeting them. They have dirty minds. The men are even worse. They ogle at you. They strip you naked with their eyes. And suppose something is wrong underneath – something is unclean. They are going to talk about you and the whole village will know.

MAN: You will make sure that everything is clean before you go out.

WOMAN: It's a risk all the same, child. Many of those men out there are full of spite. Most of them are the old man's age-mates. Many of them wanted to marry me. They felt very bitter when I chose this man. They were jealous of him too because he had a good job on the potato farms. Now that he is like this they are pleased with themselves. Don't you see, child? They willed it. Do you want me to add more sweetness to their honey?

MAN: There you go again. There must be someone to blame somewhere.

WOMAN: I am not going to invite sympathy from those who used to show us a tooth when all was well, who are sniggering now that the old man is dead.

MAN: There is still life in those bones, ma.

WOMAN: To me he is dead, child. He died sixteen years ago at the hands of those cruel beasts on the farm.

MAN: But he eats, ma.

WOMAN: When I force food into his mouth. Might as well be forcing it into a hole in the ground.

MAN: He shits too. That's a sign of life.

WOMAN: The excrement comes out of its own accord. I just collect it and throw it out at night.

MAN: And you hear his cries – like the cries of an abandoned baby.

WOMAN: We do hear voices from the dead. They communicate with us.

MAN: The old man has become an ancestor in his lifetime.

WOMAN: Child, I am beginning to be suspicious of you. Are you sure you haven't joined the neighbours against us?

MAN: I have joined them in doing nothing against you.

WOMAN: You must remember we gave you our only daughter.

MAN: You can't blame me for that. I was only ten when that happened, and she was barely six months old.

WOMAN: We have always loved your family. Although the old man had been a little disappointed that the child was a girl, he was happy when I suggested she was going to marry you.

MAN: At least I should have been consulted.

WOMAN: We knew at once that Nontobeko was meant for you. That is why we went to *abakhozi* – your parents – and made a deal.

MAN: And from that day I was engaged to be married.

WOMAN: We even finished the matters of *ukhazi* on that day.

ukhazi: the agenda or arrangements

MAN: And a great feast in celebration of the engagement was made.

WOMAN: There was no time for feasting. The old man had only been given two days off to come and show his relatives in this village the baby girl, and to perform the necessary rites for the child. After that he had to go back to the farm.

MAN: How many heads of cattle were my parents supposed to pay?

WOMAN: I do not remember well. But I know that it was more than the usual number because the old man said he was not only giving you a wife, but also a position on the farm.

MAN: A position? As his right-hand man perhaps?

WOMAN: To inherit his position as a *baas-boy*.

MAN: *Mna? Baas-boy? Hayi ... hayi*. Don't say that, mama.

mna: me
hayi: no

WOMAN: He had it all worked out. At the age of fifteen you would go and live with us so as to learn the work. Like everyone else you would start as a farm labourer and you would then work your way up to take over from him – with his help of course.

MAN: Rotten little schemer.

WOMAN: It was for your own good – and for Nontobeko's.

MAN: Had to drag me down the drain with him for the sake of precious Nontobeko.

WOMAN: You know, I fail to understand you sometimes, child. It sounds like a lot of nonsense to me.

MAN: You are damn right. A lot of nonsense.

WOMAN: Well, you can tell him all that. I am sure he will understand. It makes no sense to me. *(She goes out)*

MAN *(laughs at OLD MAN)*: So you wanted me to inherit your kingdom, did you? You wanted to take me from my parents, implant in my mind the art of slave-driving? *(He laughs again)* You know old man, I wonder what goes on in your mind. I wonder what you see there, staring at nothing. I know. Prisoners. That's what you see. Marching and singing. Digging potatoes with their fingers. Shitting right there in the fields to make manure for the potatoes. Attacking. Yes, attacking! *(He laughs for a long time)* And you nearly succeeded, too. Making me an *induna*. You nearly destroyed me. The way they destroyed you. *(Pause)* I was only twelve when this happened to you. I didn't give you the chance. I beat you by a distance of three years. Only three years before you would have taken me to your farm as your heir-apparent. Then boom! Those prisoners saved me. *(He looks at OLD MAN, then tries to taunt him by making rude gestures)* I suppose you don't know what that means. Poor old lady. She has to put up with a gelding. Well, she gets it all in her dreams. I think someone should put a video-tape into her head before she sleeps. It would be interesting to play it for her the next morning. Keep it for posterity on the days when she has had good dreams. Erase it when things go awry. *(Tiptoes to the door through which WOMAN went, peeps and goes back OLD MAN)* Let me whisper it to you, old man. *Ndiyahamba mna.* I am leaving you two to see for yourselves. Leaving you forever. We'll see who will fetch your disability pension money from the government offices. The old lady will have to go out and get it herself. She will go to the shops to buy mealie-meal, candles and

gelding: a castrated male horse

Ndiyahamba mna: I am going

paraffin. Then life will begin to change for her. She will know that the world didn't stop sixteen years ago. She will take off those mourning clothes, and mix with people. She will have no choice. She will stop having her dreams and, most of all, she will accept that Nontobeko exists only in her imagination. You can tell her when she comes that Nontobeko is dead, and I am packing my bags to transform my own dreams into reality. Of course, you are dumb. Right. I am going to tell her myself when she comes.

WOMAN enters, but MAN does not see her.

I am going to tell her, 'Goodbye ma, and your dreams. I am leaving the village for the north to join those men who are dying in order to save us. I am doing this for you ma, for the old man, and for Nontobeko in her grave.'

WOMAN: Is that so?

MAN: Yes.

WOMAN: Why?

MAN: Because I have to.

WOMAN: Is that all you can say – because you have to?

MAN: They want more men up in the north. War is brewing, and the secret recruiters have been to the village.

WOMAN: Are you taking Nontobeko with you?

MAN: You will learn to go to the well to draw water like the other women.

WOMAN: Are you taking Nontobeko with you?

MAN: You will know that there is more to life than brooding over a child who died long ago.

WOMAN: Are you taking Nontobeko with you?

MAN: Ma, I am leaving you and this is the end of your dreams. The first step is to get it into your head – accept it – that Nontobeko is dead.

WOMAN *(in bewilderment)*: Nontobeko is dead?

MAN: Yes, and you know it. Dammit ma, you can't afford to live in a dream anymore. War is brewing up north and it is not a war of dreams. You are on your own now, and as I told you the first step is to accept that Nontobeko is dead.

WOMAN: Nontobeko is dead.

MAN: Nontobeko died when she was still a baby.

WOMAN: Nontobeko died when she was still a baby.

MAN: Nontobeko was burnt to cinders.

WOMAN: *(the scene dawns on the woman, and her face becomes a contortion of agony)* Nontobeko was burnt to cinders in the hut that Jan van Wyk built us. The trouble started when they brought prisoners to work on the farm. The labourers have always been cheeky, but those *blou-baadjies* were the most defiant bastards I have ever seen. Not that they refused to work. They wouldn't dare with all those armed Boer warders. Oh, yes, they worked from dawn till sunset. Those guards did not play with them.

> blou-baadjies: people in blue uniforms, i.e., prisoners

MAN *(mimicking a warder)*: Janfek *kom ma-a-an*! Silas, Duiker *tshona*!

> kom ma-a-an: come man tshona: a command, meaning 'come on, work'

WOMAN: It's their attitude towards the old man that was wrong. You see, the guards used to leave them under the care of the old man whilst they sat on the verandah of Van Wyk's *huis* drinking brandy. The old man had always been firm with the workers, and he was not different with the prisoners. In fact his arm was even stronger when it came to them.

MAN mimes working prisoners. He is digging potatoes with his fingers and looks very tired.

They had this sneering attitude towards him, and they would whisper among themselves whenever he barked his commands. He reported this to their guards who then gave him a whip. '*Wena betha lo mabantinti nduna,*' they said. He was now armed with a whip, and whenever they started their sneering tricks they would feel his wrath on their backs.

> Wena betha lo mabantinti nduna: Fanagalo for 'You hit those prisoners, headman'.

MAN writhes with pain.

One morning the old man was in his moods again. In their defence they said he was particularly harsh on them that day. When the truckload of convicts arrived he took them straight to the fields, even before they had had their usual pee. Yes, they liked pissing, those ones. Before work they must piss. It's their excuse to steal a few minutes from the day's working time.

MAN: *Hau, hau! Asichamanga namhlanje?*

WOMAN: *Voetsek* Janfek! Move on! *(She flogs his back with an imaginary whip and he winces with pain. She drives him around the stage wielding her whip, then he kneels down and starts digging the potatoes).*

Throughout that morning the convicts worked without any rest. The Boer warders sat on the verandah of the *huis* drinking liquor. From time to time they came to the field to egg the old man on. Every time he wielded his whip they cheered; when the prisoners winced with pain they went into a great frenzy, and pride swelled in the chest of the old man. He had the prisoners in his hands – more power than he had ever had before – and he was enjoying it. And the warders were enjoying themselves, because they had someone they could trust, someone who could do what even they couldn't do – make those cheeky *blou-baadjies* work. So they spent most of their time in the *huis*. Van Wyk was happy because the work was going much faster than it used to go when he was using farm labourers. His harvest would be greater, and the whole farm smelt of prosperity that year. The missis gave me more of her old dresses than she had ever given me before. Oh yes, that was indeed a good year, and for the rest of the morning the prisoners worked. At midday things began to happen.

MAN *(standing up)*: *Haaikona.*

WOMAN: Where do you think you are going?

MAN: *Nditshiswa ngumchamo.* I am going to pass water.

WOMAN: *Voetsek.* Get down to work. *(She wields the whip. The man gets down to work.)* That was Janfek. The cheekiest of them all. The one with the dirtiest look. Then they began to sing. During their first days on the farm they always used to accompany their work with a song. But the old man stopped that long ago because all their songs were directed

Asichamanga namhlanje:
we didn't pee today.

voetsek: deragatory term for 'go away'.

haaikona: no

Nditshiswa ngumchamo:
the pee is burning me

at him. Their aim was to prick him with words. So he had stopped their singing. But ah, that day they burst into song. *(She sings)* 'Senzenina ma Afrika ...'

MAN *joins her and they both sing:*

What have we done
What have we done Africa
They hit us
The hit us Africa
They kill us
They kill us Africa
What have we done Africa

Senzenina
Senzenina maAfrika
Besibetha nje
Besibetha nje maAfrika
Besibulala nje
Besibulala nje maAfrika
Senzenina maAfrika

The old man boiled with anger. He lashed left and right with his trusty whip. The *blou-baadjies* rose. Three of them leading the mob. Silas, Duiker and Janfek. They attacked him, beat him up and left him for dead. Then they marched to our hut – still singing their song.

MAN: They had now tasted blood. They wanted more blood.

WOMAN: They set our beautiful hut on fire.

MAN: And whilst the flames rose they danced around it.

WOMAN: Singing their song of death.

MAN: *Senzenina maAfrika.*

WOMAN *(contortions of agony returning to her face)*: And Nontobeko was sleeping in the hut.

MAN: Little Nontobeko.

WOMAN: I was in my vegetable patch when I heard screams. I rushed to the hut, where I found a crowd of people fighting a losing battle against the flames. The prisoners were no longer there. Only their song could be heard from a distance, for they were now marching towards the *huis*. I tried to push through to save my baby but the flames were raging with great anger. The men pulled me back. *(She screams)* My baby! Leave me alone! I must save Nontobeko! *Ncedani bo!*

Ncedani bo: Help us please

MAN *(rushes to the WOMAN, holds and violently shakes her)* We have done away with crying, ma. We are finished with weeping.

Black people don't cry, ma. They don't weep. Do you realize, ma, he was on their side? *(Pointing at* OLD MAN*)*

WOMAN: He was only doing his work, my child.

MAN: They always say that. Every one of them. Black officials of
this regime ... civil servants who carry out the repressive
laws ... chiefs ... policemen ... in Soweto ... Langa ... New
Brighton ... throughout the land ... They are doing their
duty, whilst they mow down peaceful children marching
down the streets, voicing their hatred of the evils done to us
by those who have given themselves the divine right to
decide our destiny. That is why I am going, ma. Because
now I know that our salvation lies only in ourselves ... in
our guns.

WOMAN *(resignedly)*: My child had to pay the price.

MAN: And how much did you get out of it? What did he *(pointing at* OLD MAN*)* get besides being paralysed for life?

WOMAN: The pension money which you fetch us from the government offices at the end of every month.

MAN: Seven rands per month. Disability fund set up by Mr Jan
van Wyk who always had the interests of his workers at
heart – especially the *induna*.

WOMAN: They paid the price too. Silas, Duiker and Janfek. They
were hanged.

MAN: Not before they had burnt Van Wyk's house to ashes.

WOMAN: Yes. After burning our hut – and our child – they
marched to the *huis*.

MAN: Once you have started something you must end it.

WOMAN: Immediately the guards saw that there was a rebellion
they phoned for reinforcements who came just when the
blou-baadjies were enjoying themselves, dancing around
the burning *huis*. They were quickly rounded up, loaded
into kwela-kwelas and taken to prison.   kwela-kwelas: police vans

MAN: They were tried and three of them were hanged.

WOMAN: Silas, Duiker and Janfek.

MAN *(suddenly)*: I must go, ma. I must go to the north.

WOMAN: Do you know what you are saying? Do you know what
you are up against?

MAN: A war of freedom is never lost. It is a just war and when people fight for the cause of justice their will to continue is indestructible.

WOMAN: You have set your heart to go; go then, child.

MAN: Yes, ma. Thank you for your blessings. I am going there to fight for you, ma, and for the old man, and for Nontobeko, and for Silas, and for Duiker, and for Janfek. That is why I am going, ma.

For the first time the eyes of the OLD MAN *brighten up, and his face is full of life. He smiles broadly. They both look at him in amazement. He turns his face and looks at them.*

WOMAN *(greatly moved)*: Go, child. Quick! The war is waiting!

MAN: Remain well, ma. Remain well, old man. *(He goes out)*

END

ONE

READING

TEXTS

TO BEGIN, WE are faced with a text. Because we recognize that it is a literary text, we treat it in a special way. We try to make sense of it, to find out its possible meanings. This generally involves reading the text closely in order to understand how its components work together to reveal something of significance.

We don't do this to all texts. We wouldn't carefully examine a shopping list or a telephone directory in the hope that they might reveal a deeper significance. Nor do we necessarily treat all literary texts with the same close attention. We might race through a detective story or a light romance without being aware of the tactics that we need to employ to make sense of them. In an academic context, though, we tend to read meticulously, with the expectation that attention to detail will give us insight into several aspects of the text, including its content, form and structure, characterization and themes. Understanding texts depends on the reader's ability to interpret the signs that are being communicated.

What are these signs and how do we read them? In the case of literary texts, the most basic signs are the words on the page and the ideas and actions that they denote. But these words, ideas and actions do not merely exist in a vacuum. Rather, they occur in a specific setting. They are expressed or enacted by particular characters. And they give rise to particular ideas or effects, which literary critics usually call 'themes'.

Let's consider these textual signs in more detail.

Setting

Setting is an important feature of most literary texts. When we refer to setting, particularly in relation to a play, we are talking about two related things. Firstly, we use the word 'setting' to describe the specific time and location in which the events about which we are reading take place. Secondly, the word refers to the wider social and political context that surrounds this specific location. Take, for example, Shakespeare's *Macbeth*. The various scenes that make up the play each have their own particular location and take place at a specified time. The first scene of the play, for instance, is set on a wind-swept heath. The second takes place in King Duncan's castle. Later scenes are set in Macbeth's castle, in the court of the King of England and in a field near Dunsinane. Some scenes take place at night and some during the day. Seventeen years pass between the first and the final scene. The time and location in which these individual scenes take place is what we mean by the first use of 'setting'. The play as a whole, though, is set in Scotland around the years 1040 to 1057. This is the second sense in which we refer to 'setting'.

The stage directions that open *Dark Voices Ring* tell us that the action takes place 'Inside a hut' (p. 272). We can work out that this hut is located in a rural village somewhere in South Africa and that the action takes place in the past. We can tell this by looking at:

○ **the names of the characters:** 'Nontobeko' and 'Baas Jan van Wyk' are names that automatically suggest a South African setting;

- WHERE IS *Dark Voices Ring* set?

- WHAT WERE the clues that you used to find out the setting?

- **the language that is used:** Words such as 'Yebo nkosi', 'morena', 'huis' and 'voetsek' are specific to South Africa where English, Xhosa and Afrikaans are spoken, amongst other languages;
- **specific references to geographic locations:** 'Soweto', 'Langa' and 'New Brighton' – black townships in South Africa are mentioned;
- **the nature of the action:** The racist and oppressive relationship between the workers and the farm owner, as well as the Man's references to the 'repressive laws' of the regime and to joining 'a war of freedom', show that the action takes place during the apartheid era prior to the election of a democratic, non-racial government in 1994;
- **the publication and performance details:** *Dark Voices Ring* was first performed in Cape Town in 1979 and published by Ravan Press, a South African company. In all probability the play is set in South Africa during or before 1979.

Setting is important because it involves deliberate choices on the part of writers. We are forced to ask why they have chosen to narrate the events of the play or novel in a particular location and time. The setting provides a backdrop against which we read these events. We make sense of the events by relating them to their setting.

- Why do you think Mda sets *Dark Voices Ring* inside a bare hut?
- The hut may have a 'deeper' symbolic meaning. What might it suggest about the old woman and her relationship to the real world?

Interestingly, although the play is set in a hut and the interaction that we see is confined to a dialogue between an old woman and a young man, the central events take place some sixteen years before the scene we are witnessing. This is called a flashback and it allows the playwright to suggest that the oppression the play represents has taken place over a long period of history. With few resources and in a very short time, Mda is able to recreate a situation that is far more complex than a conversation between two people. The play is not only about Man and Woman, but about a whole range of characters, events and responses to racial oppression. Cleverly, through Old Woman and Man's recollections of the past, the setting of the play is extended out from the confines of the hut to the conditions on Van Wyk's farm and into wider society.

○ Try to name other plays or novels that use flashbacks. What purpose do flashbacks serve in these texts?
○ Think about the differences between staging a play and making a film. Why do plays have to rely more heavily on dialogue, suggestion and the imagination for their effect? Describe the settings you would choose if you were a director filming *Dark Voices Ring*.

 Character

Characterization is an extremely important aspect of any play. Unlike a novelist, a playwright cannot provide long written descriptions of events and their deeper significance. Instead, meaning has to be conveyed through what the characters say and do. When we watch or read a play we are usually interested in answering these questions:

1 Who appears (on the stage or on the page)?
2 What are the characters like physically?
3 What personality traits do the characters have?
4 What ideas do they have or represent?
5 How are their identities and personalities established?
6 What connection is there between a character's personality and his or her behaviour?
7 What purpose does a character serve in conveying meaning?

In a play, many of the answers to these questions will be found in what the characters say – in their dialogue. The reader or audience also looks at how they behave and takes account of non-verbal factors such as their physical appearance, their body language, their habits and the clothes they wear. Usually, playwrights guide their audiences towards particular understandings. The stage directions at the beginning of *Dark Voices Ring*, for example, tell us quite a lot about the three characters in the play.

Imagine that you have been asked to direct a production of *Dark Voices Ring*. You have to choose the actors who are going to play the various parts, decide on what they are going to wear, and give them advice on how they should act on stage and speak their lines.

Here are some questions to help you:

✪ What physical appearance would you have in mind when casting the roles of Woman, Man and Old Man?

✪ You have been told how Woman and Old Man are dressed, but what sort of clothing would you choose for Man?

✪ How would Woman and Man speak?

✪ Obviously, apart from Old Man, who remains immobile, the characters need to move and do things during the course of the play. They can't just talk to each other. Sometimes the playwright describes their actions. Mostly, the decision is left up to the director. Read through the text noting in pencil in the margin what you want the characters to do at different points in the dialogue.

Characterization in a play often emerges slowly. The personalities of the central characters are built up gradually as the action develops. Central characters (main characters) may have complex personalities made up of several traits, attitudes or beliefs. Characters may also undergo changes, with different aspects of their personality or temperament emerging in shifting circumstances. When we read a play, we should not judge a character on the basis of a single action, but against the evidence of the play as a whole. Indeed the meaning of the play is likely to hinge not only on the differences that exist between characters, but on conflict between different aspects of a character's personality. Secondary characters (those who play smaller roles) may also serve an important function. Often they represent a particular viewpoint or highlight the significance of the actions or words of the central characters. Minor characters in a play are often based on recognizable stereotypes – for example, the clown, the villain, the dreamer, the outcast and so on. The use of stereotypes frees the writer from having to develop the identity and significance of the character through dialogue.

Dark Voices Ring is a very short play and it has only three characters. Accordingly, it is probably inappropriate to expect that
the characters will be developed in any detail. This doesn't mean
that the characters' words and actions are insignificant, or that
their personalities do not undergo change, or that interpersonal
conflict is absent. Let's look for a moment at part of the dialogue
from near the beginning of the play. Read the section on pp.
273–274 from 'MAN: The old man says he wants to see
Nontobeko?' until 'MAN: You can dream of the good times ...
when Nontobeko was born. It must have been a happy event for
you and the old man.'

✪ How would you describe the relationships between the
  characters?
✪ What does the dialogue tell us about Woman's character?
✪ What do Man's responses reveal about his attitude to the
  Woman?
✪ Based on the extract above, how would you describe the
  character of Man?
✪ According to the extract, why does Old Man no longer speak?

As the action of the play unfolds, we become aware of other
dimensions. Compare and contrast the two versions of
Nontobeko's birth in the extract on pp. 276–277. Read from
'WOMAN: They say I was found lying there by passing women
who were also taking food to the farm labourers.' until 'WOMAN:
... watching Nontobeko's struggle to break free from the dark
prison of the womb ... laughing their lungs out ... whilst I sleep
there ... unconscious.'

✪ Which version of Nontobeko's birth do you think is most
  accurate? Why?
✪ What does the discrepancy between the two versions suggest
  about the Woman?

By comparing and contrasting parts of the play, we can produce
a detailed picture of the development of individual characters.

Even if certain characters seldom appear or say anything, often we are able to discern what they are like from what other characters say about them.

✪ Use your journal to make notes about the following characters from *Dark Voices Ring*: Old Man, Man and Baas Jan van Wyk. Describe each person's age, class, position and personality.

It is interesting that the characters who actually appear on stage are not given personal names, while other people who feature in the story are given names – Nontobeko, Baas Jan van Wyk, Silas, Duiker and Janfek.

✪ Why doesn't the playwright name his central characters?

 ## Plot and Theme

The plot of a play is the main events that occur. It is the answer to the question: what happens? In a short play it is fairly easy to map out the plot in some detail, but in a five-act play this is more difficult to do. In these cases we have to pick out those sequences of action that seem most significant.

✪ Make a list of the most significant events in *Dark Voices Ring* in order.

When dealing with plot, we need to be alert to when and how information is revealed to the reader or audience. Suspense is an important aspect of many literary texts. The writer deliberately withholds information from either the reader or from other characters. This allows tension to build up. You will have noticed that for over half of *Dark Voices Ring* we are not aware that Nontobeko is dead. The point at which this is revealed is a pivotal moment. The entire presentation of the play balances on this revelation.

POWER

Look at the dialogue up until the point (on p. 281) where Nontobeko's death is revealed.

✪ What is the main topic of conversation?
✪ What is the mood or tone of this section of the play?

Now examine the dialogue after this revelation (from pp. 281–286).

✪ How does the focus of the play shift?
✪ What is the mood or tone of the latter part of the play?
✪ In your opinion, which of the two parts of the play is more powerful in its effect?

I expect that you found that the play became more powerful as it moved towards its end. We call the moment of greatest tension or conflict the 'climax'.

✪ For you, what is the climax of *Dark Voices Ring*?

When we talk about the theme of a play or novel we are talking about something different from its plot. A theme is one of the main topics or ideas that is conveyed by the combination of setting, character, dialogue and plot. It is an abstract statement concerning the central ideas of the text.

✪ Try to formulate a statement that captures the central ideas conveyed by *Dark Voices Ring*.

Theme and meaning are seldom expressed directly in texts. Part of the pleasure of reading a literary text is the work that we have to do to interpret it. Texts in which everything is made explicit, in which all meaning is on the surface, offer less satisfaction because they do not require our engagement.

In English literature, irony is one of the most important literary devices used to involve the reader or viewer in the meaning-making process. Irony results from a discrepancy or gap between what is thought, felt, said or seen and the reality of the situation. It always involves differences in perception. Sometimes the difference in perception is between characters. One person sees and interprets something in a manner that another character or the reader knows is untrue. On other occasions, there is ironic discrepancy because characters claim to hold a particular idea or position, but we as readers know this is untrue. At its most subtle, the irony may be between the idea presented in a novel or play and what we know to be real. For instance, to use a crude example, a novel that depicted South Africa's urban informal settlements as residential paradises in which a happy population lives in perfect harmony would be drawing on the reader's knowledge of the discrepancy between this representation and the reality of daily life.

IRONY IN ACTION

One of the most notable ironic speeches is William Shakespeare's *Julius Caesar* (first performed in 1599, but only published in 1627).

Julius Caesar, a powerful Roman Senator, has been murdered by Brutus, who now seeks power for himself. After Brutus has addressed the Forum, Mark Antony — Brutus's opponent — arrives with the dead body of Caesar. Let's eavesdrop on his speech:

Friends, Romans, countrymen, lend me your ears;
I come to bury Caesar, not to praise him.
The evil that men do lives after them,
Their good is oft interred with their bones;
So let it be with Caesar. The noble Brutus
Hath told you Caesar was ambitious.
And if it were so, it was a grievous fault,
And grievously hath Caesar answer'd it.
Here, under leave of Brutus and the rest,
(For Brutus is an honourable man,
So are they all honourable men)
Come I to speak in Caesar's funeral.
He was my friend, faithful and just to me;
But Brutus says he was ambitious,
And Brutus is an honourable man.
He hath brought many captives home to Rome,
Whose ransoms did the general coffers fill:
Did this in Caesar seem ambitious?
When that the poor have cried, Caesar hath wept;
Ambition should be made of sterner stuff:
Yet Brutus says he was ambitious,
And Brutus is an honourable man.
You all did see on the Lupercal
I thrice presented him a kingly crown,
Which he did thrice refuse. Was this ambition?
Yet Brutus says he was ambitious,
And sure he is an honourable man.
I speak not to disprove what Brutus spoke,
But here I am to speak what I do know...

(Act 3, Scene 2, Lines 75–103)

- Having read this extract, do you think Antony's claim 'I come to bury Caesar, not to praise him' is ironic? Why?
- Why is Antony's reference to Brutus as 'honourable' ironic?
- What effect does the repetition of 'And Brutus is an honourable man' have?
- Why do you think Antony brings Caesar's dead body with him?
- What message do you think the speech conveys about Caesar and Brutus respectively?
- What makes this an effective speech?

There are several instances of irony in *Dark Voices Ring*. For instance, you might want to look at Woman's pride in her former status and contrast this to the reality of the subservient position that she and Old Man actually held. You also might find it ironic that the prison labourers working on the farm lashed out at Old Man and his family, when the real system that oppresses them is represented by Baas van Wyk, the warders and the political system that allows prisoners to be used as slaves.

To conclude this discussion of theme, look at the final moments of the play. Read from 'MAN: (*suddenly*) I must go, ma. I must go to the north' (p. 285) until the end of the play.

- What do you think of the play's ending?
- What does the behaviour of the Old Man and the end of the play suggest?

By now you should be in a position to talk about how this play works and what it means. Having grasped its basic elements, you could go on to look more closely at some of its textual details. For example, you could analyse Mda's choice of words (the *lexis* of the play) or the grammatical structures of his sentences (the patterns of *syntax*). You could also exam-

Nomhle Nkonyeni (Woman) and Nkos'omzi Ngcukana (Man) in *Dark Voices Ring*.

ine how the play draws on certain patterns of imagery and discuss its symbolic value. Entire books have been written on imagery in Shakespeare and syntactical patterns in the plays of Harold Pinter and Tom Stoppard. All these analytical adventures would keep you focused on the text. If you doubt your own critical abilities or if you are dealing with a particularly difficult text, you may want to consult other texts that discuss the play, poem or novel that you are dealing with currently.

Here is an extract from a critical article. Read it carefully.

'People are being murdered here'

by Andrew Horn

In *Dark Voices Ring* there is a return to the specific conditions prevailing in South Africa. With muck-raking intensity, the play sets out to expose as Mda notes, 'the ill-treatment of farm labourers' in the Republic. An old couple live in isolation in a meanly appointed hut, somewhere in the rural *platteland*. The husband is a catatonic; the wife, a nearly paranoiac recluse, a fugitive from a social censure which, paradoxically, she refuses to acknowledge. The play's single scene is built on an impassioned and argumentative duologue between the woman and the young man who she believes to be the husband of her daughter Nontobeko. It is revealed gradually that the Old Man has been a *'baas-boy'* – a black overseer – on a remote white-owned potato farm which relied upon, and grossly abused, convict labour.

> Throughout that morning the convicts worked without any rest. The Boer warders sat on the verandah of the huis drinking liquor. From time to time they came to the field to egg the old man on. Every time he wielded his whip they cheered; when the prisoners winced with pain they went into a great frenzy, and pride swelled in the chest of the old man. He had the prisoners in his hands – more power than he had ever had before – and he was enjoying it. And

the warders were enjoying themselves, because they had someone they could trust, someone who could do what even they couldn't do – make those cheeky *blou-baadjies* work.

As reward for his cruel reliability, the Old Man and his family have enjoyed comforts and privileges denied the other black workers on the farm – better accommodation, a more substantial diet, occasional gifts of cast-off clothing from the farm owner:

> WOMAN: Van Wyk was giving him a bag of mealie-meal every month. We had our own vegetable patch, our own hut, and our own milk cow. Besides, all the other farm labourers were under the old man's command. Kaptein, they called him. When he called them they all used to answer: *Yebo nkosi ... ewe nkosi ... morena.* The master of the farm called him 'my faithful *induna*'. The old man wanted his son to inherit his position.
>
> MAN: To inherit the power and the glory and the kingdom of Baas Jan van Wyk's farm.

But when the prisoners mutiny, they turn first on the hated 'baas boy', beating him unconscious, burning his home and in it the infant Nontobeko, before setting fire to Van Wyk's house.

Painfully, the Young Man leads the Woman to an understanding and an acceptance of what happened to her so many years earlier, until, finally, not only does she concede the reasons for the collapse of her life but gives her blessings to the Young Man's wish to join the anti-government guerrilla forces. Even the immobile Old Man, who had presumably overheard all, is roused from his apathy and paralysis to signal a mute affirmation of the Young Man's resolution...

But *Dark Voices Ring* is not primarily the story of this young intended guerrilla, whose plans have, anyway, been confirmed well before the play's action commences. Rather, it is the Old Woman and her complex responses to experience which form the core of the text. Withdrawn, estranged from others, she has transmuted her repressed social anxieties into erotic fears...

What is unclean in her life is not, however, sexual. It is the festering and unacknowledged guilt of betrayal. In her paranoiac isolation, she hopes to preserve the fragile struts of her obsessively constructed fantasy, to fend off the insurgency of reality. ... she refuses to accept both what has happened and why it has happened ... The social and material prerequisites of being the wife of the *induna* have fortified the Woman's belief in the legitimacy of her life on Van Wyk's farm. Dissonance, however, has been introduced by the events of the workers' insurrection. She has attempted to resolve this tension by means of defensive misperception and an avoidance of potential sources of dissonant information. It is more stabilizing, causes less anxiety, for her to deny the fact of her daughter's death than to accept that the ultimate responsibility for the death lies with herself and her husband and the social roles they chose to adopt. But the Young Man re-stimulates the dissonance by deluging the Woman with material antagonistic to her fantasies and increasingly difficult to confute. By making her re-live the events she has laboured imaginatively to reassemble, the Man achieves an imbalance, between her fictive memory and that reluctantly recalled truth, so great that she is led ultimately to a therapeutic resolution, the revision of her now untenable view of herself ...

In rehearsing the birth of Nontobeko and the rebellion of the convict-labourers, the Woman is able to reintegrate her personality and, the suggestion is, to re-enter the world of other people ...

This realization [of the death of Nontobeko] serves dual purposes in the play, for while it represents the beginning of emotional renewal and social rehabilitation for the individual character, it also serves as a warning to those in the audience who may be tempted to accept in perhaps more subtle forms, the role of 'baas-boy' in their own spheres of activity.

(FROM: Zakes Mda, 1990. *The Plays of Zakes Mda*, pp. xx–xxvii. Johannesburg: Ravan.)

✪ What insights have you gained into *Dark Voices Ring* from reading Andrew Horn's article?

✪ Make a list of questions that you have after reading the article. Discuss these questions with another student or with your lecturer or tutor.

⬤ Assumptions and Implications

Up to now we have considered *Dark Voices Ring* solely as a text. Our analysis has been based on the text and we have referred to another literary critical text in our search for a clearer understanding of the play. We need to ask ourselves whether we can form a complete and accurate understanding of a poem, play or novel on the basis of a close reading of the text alone?

These questions do not have easy answers. I am sure that after a close examination of the text most people would feel that they had understood *Dark Voices Ring*. As a text, it seems to present few difficulties. After all, we read and understand texts every day without feeling the need to bring in other, extra-textual material. In addition, much of our schooling has been based on teaching us the techniques of close reading. Isn't the critical process just a particularly intense version of what we do naturally when we read? For me, the fact that reading in this way seems natural is a good reason to examine the assumptions and implications of doing it. By this stage in *Love, power and meaning* you are probably deeply suspicious of any activity that appears to be an aspect of everyday life!

To start, let's go back to the discussions of formalism and New Criticism in *Critical conditions*. Both New Critics and formalists assume that the text itself is the object of study. The meaning of the text is in the words on the page; there is no need to consider the author's intentions, the social context or the ways in which different readers respond to the text. Beneath this claim for the autonomy of the text are two other claims. First, because the meaning of the text is embedded in the words on the page, it must remain the same across time and in different cultures. Secondly, because the meaning is in the text and is separated from factors that might influence it (personal and political preferences, for instance), it can be recovered objectively by a highly trained reader. Literary criticism, then, is not just an expression of the whims of individuals, it is a science.

✪ Do you think that texts express a meaning that people in different cultures and at different times in history will agree upon?

✪ To what extent do you think a critic's opinions, beliefs and cultural position influence his or her interpretation of a text?

✪ If literary criticism is not an objective scientific activity, how can we tell whether one reading of a work is more valid than another?

As we pointed out previously, close textual analysis is still carried out today and it has many points in its favour. No one would deny that the techniques of close reading have made us sharper readers. Traditionally, though, formalists and New Critics have been antagonistic to modern theories of literature and culture. They have tended to portray these theories as politically and ideologically biased, while claiming that there is no ulterior motive in their methodology. Yet, if we examine the origins and practices of New Criticism, particularly, another story emerges. As we have shown in *Critical conditions*, close textual analysis coincides with the emergence of English as a university discipline. It offered teachers and lecturers a convenient and manageable method for dealing with large groups of students and a wide range of literary material. Good literature was seen as literature that uttered certain truths about the human condition.

Human condition this phrase is used to refer to universal or general features of human life, in particular the experience of suffering.

Because the human condition is complex, the literature that best expressed it would also be complex. As a result, New Critics tended to value work that had a clear moral message and which expressed that message through elaborate symbolic forms. Anything too direct and too accessible had to be 'bad literature' and was therefore not worth studying.

This stance has several obvious advantages. First, it reserves the power of identifying and commenting on cultural richness for an intellectual élite. They become the custodians of the nation's cultural heritage because they alone have the capacity to read and understand high art and culture. Ordinary people who are not trained to interpret high art do not have a say; they are excluded. Second, by emphasizing the moral basis of literature, it is possible to draw attention away from its political implications. Indeed, if the poem or play has political meanings, then those may need to be obscured by focusing on matters of style or symbolic complexity so as not to cause any harm. Unless, of course, the political or moral message is one that the establishment wishes to promote.

In contrast to this denial of politics, the entire New Criticism methodology has a political meaning. Examining nothing but the text stops students and critics from experiencing the rich web of cultural texts that themselves reveal the power and oppression of people. We never have to confront any direct testimony of poverty, persecution, abuse and exploitation. Instead, we can deal exclusively in the realm of representation and abstraction. Not only does this focus on the fiction of literature potentially limit our vision and understanding, it also diminishes the possibility of our taking any direct political action. Finally, by insisting on the primacy and importance of the text, teachers and critics are able to avoid discussing their own political positions.

There is another fundamental difficulty in ignoring the social and political conditions that surround literary texts. As we have seen, contemporary literary and linguistic theories show that language and knowledge are inseparable from society. Contrary to New Critical or formalist claims that meaning is transcendent and unchanging, it seems that language can and does signify different things in different context and times. This may seem unimportant when we're dealing with a piece of recent writing from our own milieu. The language and the things that it signifies are likely to be familiar. The problem really arises when we are looking at texts written in environments, either in the past or from distant contexts, with which we are unfamiliar.

Without an acute understanding of the context, it may not be possible to provide an accurate reading of a Shakespeare play. Similarly, it may be inappropriate to impose modern ideas on earlier historical periods, like Renaissance England or Europe in the Middle Ages. Of course, it may be argued that the text itself contains all the information needed for a sensitive reading of culture. New Critics will also point out that they do read beyond the text and that they are generally well-versed in history, philosophy and other arts. Nevertheless, at best, context is used merely to fill in the background. It seldom becomes the primary focus of study. When the relationship between text and context is discussed it is usually as a means of taming the text and fixing its meaning. The possibility that the same text may mean different things across time and culture is rarely acknowledged, if ever. Finally, these critics of the text, who cast aspersions on investigating the multiple contexts in which works are written and read, often use literary texts as a means of reconstructing the social and historical past. The literary text, as a reflection of reality, becomes a point of access through which we can know what it was like to be an Elizabethan or to live under the British Raj in India. This is a highly questionable practice. Personally, I am not sure whether the full meaning of a literary text is ever recoverable, let alone whether we can reach deep insights into society merely through the imaginative products of individual writers.

My views on close reading

My own opinion of close textual analysis should be clear – I did say at the outset that I would be honest about my own views. At best, close critical readings are, I think, useful first steps. For one, I would question whether the full meaning of the play is in the text alone. I think it is important to consider context and reception and to investigate what the text says and does socially and politically. Without these dimensions, I believe that close critical readings of texts are sterile. They engage us in a futile activity that has no worth outside the academy. I can see no purpose in providing ever more elaborate interpretations of Milton's *Paradise Lost* or Wordsworth's *The Prelude*, especially in a country where poverty, homelessness and unemployment are endemic. To me, literature and literary study need to serve some definable social function. Exactly what and how, I'm not sure. What do you think?

- ✪ Do you think that the New Critics would consider *Dark Voices Ring* a good play? Why or why not?
- ✪ What sort of literature did you read at school? Was it relevant to your life and society?
- ✪ Do you think that art and literature should serve a social purpose?

. . . .

THE AUTHOR OF the text is an obvious subject for attention if we are going to go beyond just the words on the page. Perhaps by accumulating information about the author, we can gain greater insight into the text. Perhaps, too, knowing about the author's life and intentions may help us see the text from another perspective. The influence of biographical details on interpretation and evaluation was discussed at some length in the first two chapters of this book. Read through this discussion again and then read through the items in the 'Fact File' that we've prepared on Zakes Mda.

CONSIDERING THE AUTHOR

⬤ Fact File on Zanemvula Kizito Gatyeni Mda

ITEM 1: Entry in David Adey, Ridley Beeton, Michael Chapman and Ernest Pereira. 1986. *A Companion to South African English Literature*. Johannesburg: A.D. Donker.

MDA, Zanemvula (Zakes) (1948–)
Born in the Eastern Cape, he attended school near Johannesburg. In 1963 he left South Africa with his family to live

Zakes Mda

in neighbouring Lesotho where he worked for a time in a bank before becoming a school teacher. An exhibited painter and graduate in dramatic arts, in 1978 he won a Merit Award for *We Shall Sing for the Fatherland* (first performed in 1979), a tragi-comic examination of the hopes and fears of black SA veterans after a victorious War of Independence. Both *Dark Voices Ring* and *Dead End* were first staged in 1979 and *The Hill*, a play about migrant workers in Lesotho, won Mda the Amstel Playwright of the Year Award in 1979. Three of his plays comprise the volume *We Shall Sing for the Fatherland and Other Plays* (1980).

ITEM 2: An abbreviated Curriculum Vitae for Zakes Mda

1 Personal
Name:	Zanemvula Kizito Gatyeni Mda
Pen Name:	Zakes Mda
Date of Birth:	October 6, 1948
Marital Status:	Married

2 Education
1985–89	PhD, University of Cape Town.
1981–84	MFA (Theater), Ohio University, Athens, Ohio, U.S.A.
1973–76	BFA (Fine Arts) International Academy of Arts and Letters, Zurich, Switzerland.
1965–69	Cambridge Overseas School Certificate, Peka High School, Leribe, Lesotho.

3 Work Experience:
1995–Present	Full-time writer, painter, filmmaker, based in Johannesburg, South Africa.
1994–95	Visiting Professor, Wits.
1993–94	Professor, English Department, University of Vermont.
1992–93	Visiting Research Fellow, Southern African Research Program, Yale University.
1985–92	Lecturer, Senior Lecturer and Professor,

English Department, National University of Lesotho.

1984–85	Lesotho National Broadcasting Corporation: Controller of Programmes and Consultant to their television project.
1980–81	American Cultural Center (USIA), Maseru, Lesotho: Cultural Affairs Specialist.
1972–79	Ministry of Education, Lesotho.
1971–72	Clerk, Barclays Bank PLC, Leribe, Lesotho.
1984–1998	Involved in conducting writing workshops in various countries.

4 Books (creative)

2000	*Heart of Redness*, a novel (Cape Town: Oxford University Press)
1988	French edition of *Ways of Dying*, published under the title *Le Pleurer* (Paris: Editions Dapper).
1998	*Penny and Puffy*, a children's book, co-authored with Mpana Mokhoane (Reykjavik, Iceland: Aeskan).
1998	*Melville 67*, a novella for youth (Johannesburg: Vivlia Publishers).
1995	*She Plays with Darkness*, a novel (Johannesburg: Vivlia publishers).
1995	*Ways of Dying*, a novel (Cape Town: Oxford University Press).

5 Books (Academic)

1993	*When People Play People: Development Communication through Theatre.* (New Jersey & London: Zed Books, Johannesburg: Witwatersrand University Press).

6 Playscript Productions

1998	*Love Letters*, play published in *Let us Play* (Johannesburg: Vivlia publishers).

1996	*The Nun's Romantic Story* published in *Four Plays* (Johannesburg: Vivlia Publishers).
1993	*Four Works* (Johannesburg: Witwatersrand University Press).
1990	*The Plays of Zakes Mda*, with an introduction by Andrew Horn (Johannesburg: Ravan Press).
1981	*The Hill*, a full-length play in two issues of *Scenario*, Johannesburg.
1980	*We Shall Sing for the Fatherland and other Plays*, an anthology of three plays (Johannesburg: Ravan Press).

ITEM 3: **Extracts from an Interview with Zakes Mda, Maseru, 8 February 1987.**

*T*here is a problem with establishing when your plays were written. I understand that Dead End was written while you were still at school. Is that your first play?

No, I had written earlier plays but I don't know where they are now. *Dead End* was written in 1966 – it was the third play I wrote. The next was *We Shall Sing for the Fatherland* in 1975, followed by *Dark Voices Ring* (1976), *The Hill* (1978) and the plays I wrote in America, *The Road* in 1982 and *Joys of War* in 1983.

I know that your father, A.P. Mda, was very influential in the African National Congress Youth League and the Pan African Congress. To what extent have you been influenced by your father's political affiliations? Were you, for instance, nurtured on Africanist ideals?

At home, there was never any direct influence from Pan-Africanism or Africanist ideology. But as kids we could see the comings and goings of various political figures like Mandela and Sisulu. In fact at one stage my brothers and I stayed at Mandela's house in Orlando East. Therefore we acquired some form of political consciousness at an early age, which was not due to direct parental influence, but from mere observation. Later, when we were a bit older, my father used to analyse the South African situation for us. Of course, at an early age I believed in Pan-Africanism and African Nationalism. But as time has passed I have changed. I think that now, politically, I am more drawn to the so-called 'progressive' tendencies and my political inclinations are sympathetic to Marxist-Leninism. Both as a scholar and student I have found a Marxist-Leninist analysis of the South African situation, and of the arts in general, very useful.

All your plays combine incisive political comment with what may be seen as European avant-garde, modernist technique. Reviewers have likened your work – particularly We Shall Sing to the Fatherland – *to that of Samuel Beckett and Eugene Ionesco. What influence have Western dramatists had on your work?*

When I wrote *We Shall Sing for the Fatherland* I had never read any Beckett or Ionesco. I read Beckett much later and, to my utmost shame, I've hardly read anything of Ionesco. I do think my style was influenced a lot by European contemporary theatre. When I was at school I read plays by Joe Orton, Tennessee Williams and Sam Shepard. But I was also greatly influenced by two African playwrights, Athol Fugard and Wole Soyinka.

In your own writings on theatre in Momentum *and* The Classic *you state that you 'have dismally failed to respond to the strange aesthetic concepts so cherished in the Western world that profess that artistic creation is an end in itself, independent of politics and social requirements'. Yet, your plays are artistically innovative. To what extent do you see your works as departing from Western notions of art? Do you, for example, draw on the African performance modes of the oral tradition?*

I don't believe art necessarily detracts from social relevance. I also do not see how social relevance makes a work inartistic. The symbols and images I use are not necessarily drawn from African culture as such. Some of them might be, but I am not particularly conscious of the fact. My style – not my content, nor my innovations and experimentations – is based on international models, but the role that I hope to play as an artist, and the role I hope my work plays, is that of social commentator and social commentary. I am against art for art's sake – in African aesthetics that is a strange concept because the artist was actually a social commentator. However, I do not want my work to end as social commentary only. I want my theatre to be a vehicle for a critical analysis of our situation. I want it to rally people to action.

Dark Voices Ring ends with the Man leaving to join the liberation forces in the North. Thus the play seems to advocate direct political intervention in a way that your other works do not. Do you think playwrights have a responsibility to point towards such action?

My personal belief is in an armed struggle and the use of other pressure. If a playwright can point to such action he must do so.

FROM *South African Theatre Journal,* Vol. 2, No. 2, 1988.

Theatre has always played a vital role in reform and reflection; and in South Africa, a society characterized by racial segregation, economic exploitation and political oppression, it continues to be a significant voice in the resistance of the repressed majority. This it does despite the censorious nature of the environment in which the artists work. In fact the harsher the hand of the censor, the more impressive our theatre.

When I started writing plays in the mid-sixties there existed in the various townships a form of theatre that seemed to be based on a set formula. The plays, and these always incorporated a lot of song and dance, were characterized by a policeman who was normally very slow-thinking and insensitive, a wily priest, a comical school teacher, the street-wise and fast-talking tsotsi, a diviner who was sometimes but not always a fake, a shebeen-queen, and a township gossip. The plot would involve a church service, usually of the 'Zionist' denomination, a jail scene, a wedding, and a funeral. There would be plenty of weeping, exaggerated gestures and forceful speech. The different story-lines were almost always based on this formula.

The best of these works were often humorous social satires and were to a large extent relevant to the everyday experiences of the people, but they did not carry any profound political message. A few did touch on political themes, but in most cases this would be incidental. The writers' intention was entertainment and the primary interest commercial. Of course the censor also played his part to make certain that whatever political matter was contained in the plays would be that which advocated political reactionism and the maintenance of the status quo. Playwrights were required to have their scripts read by government officials who knew nothing about theatre before they could get venues for the performance of their plays in the township halls.

New forms of stylistic expression emerged in the seventies. With the advent of plays like Maqina's *Give Us This Day*, although to a large extent set within the formula, we saw playwrights consciously and openly siding with the progressive tendencies of the political struggle. We heard theatre voicing the political aspirations of the South African majority. It is at this time that groups like Workshop 71, the Theatre Council of Natal, the People's Experimental Theatre and the Serpent Players emerged. Theatre became part of the whole liberation movement which culminated in the June '76 resistance, and continues today in the struggle that is going on within and outside the country.

As with most artists of my generation, the historical developments in South Africa, including the June '76 resistance, have had a great impact on my work.

Although writing thousands of miles from the country, the characters and situations I depict in my drama continue to be motivated by social, political, economic and historical factors in South Africa. I have dismally failed to respond to the strange aesthetic concepts so cherished in the Western world that profess that artistic creation is an end in itself, independent of politics and social requirements. I draw from the traditional African aesthetics where art could not be separated from life. In our various African societies the artist was a social commentator. *Imbongi*, the praise singer, also became social critic. Among the Pende the *Mbuyo* masks were made to represent certain people in the society, and through choreography and music the performers would vent social criticism by miming certain behaviour patterns of the rulers with whom they were dissatisfied. In this way wrongs were righted. I do not believe in the universality of the human condition, for the human condition is always determined by social, political and economic factors. It therefore changes from time to time, and from place to place...

Since the transition from the township theatre to the radical one of the late seventies and eighties we have seen greater commitment on the part of both the writers and the audiences in South Africa. It is not unusual for a play like Manaka's *Egoli* to be performed not only in theatres and halls (that is, before it gets banned), but at weddings and funerals. The people want to see this tragedy at a happy event like a wedding, for as an optimistic tragedy it embodies the will of the oppressed to continue the resistance and to overcome in the end.

FROM: Daymond, M.J. et al. (eds.), 1984. *Momentum: On Recent South African Writing*. Pietermaritzburg: University of Natal Press.

❂ What insights have these four items given you into *Dark Voices Ring*?

❂ What relevance does Mda's political background have for a reading of the play?

❂ In the interview and in the piece that Mda wrote for *Momentum* he explicitly rejects Western aesthetic concepts. Use *The Oxford Advanced Learner's Dictionary* to look up the word 'aesthetic'. What does it mean? What effect do you think this rejection of Western aesthetics has had on *Dark Voices Ring*?

❂ What is your response to Mda's statement in Item Four: 'I do not believe in the universality of the human condition, for the human condition is always determined by social, political and economic factors'?

✪ Do you think that *Dark Voices Ring* has the capacity to talk to and motivate ordinary people in the way that Mda would like his works to? Why?

 Assumptions and Implications

Assuming a connection between a writer's life and his or her work is not far-fetched. Clearly, books are written by people who have a particular story to tell and it is reasonable to think that the views, attitudes and experiences of these people may have had an effect on the story.

✪ Find out about the life and views of a writer whose work you have read recently. You could consult dictionaries of literary biography, interviews, biographies or autobiographies and the author's own critical writings. The library or the Internet may be useful.

The difficulty arises when we assume that there is a direct link between the author's life and the literary text that he or she has produced. Literary works are usually imaginative representations of the world. Although they may deal with real events and actual conditions, they do not necessarily represent the author's personal experience or a factual version of the things that they deal with.

Consider, too, the difficulty that may arise between the author's claims for a book and how it is generally interpreted. For example, what if Mda had claimed (and he hasn't) that *Dark Voices Ring* was intended to discourage oppressed workers from seeking political freedom? As readers, we disagree vehemently with this reading of the play. Would we be likely to change our reading of the text so that it fitted the author's intentions? These points will be familiar to you from our previous discussion. Rather than going over this ground again, I want to consider the issue of the author's political beliefs directly and specifically.

In Mda's case, his political affiliations are well known. He has been an active campaigner against apartheid and for a non-racial democracy. All his works are concerned with black political emancipation. With these political credentials, we might feel pleased that we have had a chance to study his writing. But what if he wasn't any of these things? What if he was an apologist for apartheid? Should this make any difference to whether we read and value his writing or not?

This is a difficult issue to resolve. Several writers who have been (or still are) considered 'great' have held political views that are today unpopular or unacceptable. T.S. Eliot (1888–1965), for example, is considered to be one of the most influential poets of the twentieth century. His poems are studied in schools and he features on university syllabuses. Yet, his political beliefs were extremely conservative and he was an admirer of fascism, the political ideology of Nazi Germany. Roy Campbell, the South African poet, also defended fascism and has been accused of having racist views. Students of English literature could expose the political skeletons in many famous writers' closets. If we did this, we would not be the first or the last to do so. Governments and institutions all around the world have suppressed and continue to suppress literary works and to banish or jail writers for expressing unpopular political ideas.

I find acts of repression reprehensible. I am opposed to censorship of any

THE BALLAD OF ENGLISH
LITERATURE
(*to the tune of 'Land of Hope and Glory'*)

Chaucer was a class traitor
Shakespeare hated the mob
Donne sold out a bit later
Sidney was a nob

Marlowe was an elitist
Ben Jonson was much the same
Bunyan was a defeatist
Dryden played the game

There's a sniff of reaction
About Alexander Pope
Sam Johnson was a Tory
And Walter Scott a dope

Coleridge was a right winger
Keats was middle class
Wordsworth was a cringer
But William Blake was a gas

Dickens was a reformist
Tennyson was a blue
Disreali was mostly pissed
And nothing that Trollope said was true

Willy Yeats was a fascist
So were Eliot and Pound
Lawrence was a sexist
Virginia Woolf was unsound

There are only three names
To be plucked from this dismal set
Milton Blake and Shelley
Will smash the ruling class yet

Milton Blake and Shelley
Will smash the ruling class yet.

(FROM: Terry Eagleton, 1986. *Against the Grain: Selected Essays*, p. 185. London: Verso.)

type. At the same time, I feel uncomfortable about supporting writers who advocate fascism, sexism, racism or homophobia. This leaves me in a quandary. If I decide never to read Eliot again, I might be depriving myself of a rich and rewarding experience. I might also cut myself off from writing that is essential to academic literacy. But if I do read him, I might tacitly be supporting his conservatism. Moreover, where is the line drawn and who draws it? If I am personally opposed to communism, would this be a legitimate reason for avoiding the works of communist writers? And why stop at communists?

✪ Do you think that the author's political beliefs should influence our response to a literary work?
✪ Do you think there is a danger that if we approve of the personal political beliefs of a writer, we might be more likely to view his or her work favourably?
✪ How do you think we should deal with literary works that express ideas about politics, religion or sex that offend us or with which we disagree?

. . . .

CONTEXT

FROM READING *Dark Voices Ring* and the biographical material on Mda we know that his work is deeply concerned with social and political issues in African and South African society. An analysis, then, that didn't take context into account would seem ill-informed.

Criticism that places great emphasis on the relationship between literature and society usually falls into one of two schools of thought – Marxism or historical hermeneutics. Both of these have been dealt with briefly in Chapter one and I have also discussed Marxism in the previous chapter, so I won't go into them here. Rather, I want to ask you to draw on your knowledge of political events in South Africa between 1948 and 1980, before presenting you with three historical documents to consider.

POWER

Fact File on South Africa: 1948–1980

While you may be familiar with the history of oppression and resistance in South Africa, you may not be aware of the importance theatre assumed during the struggle for liberation. This is mentioned in Item Three of our fact file on Mda, but you might like to

If your knowledge of South African history is shaky, you may want to consult T.R.H. Davenport and Christopher Saunders, 1999. *South Africa: A Modern History*, 5th edition. Basingstoke: Macmillan.

know that during the early 1970s, the Black Consciousness movement adopted an aggressive cultural and theatrical policy in order to promote black liberation. Item One is taken from the unpublished manuscript of a Black Consciousness speech by Saths Cooper, which was delivered at the Mofolo Hall in Soweto on 21 January 1973.

ITEM 1: 'What is Black Theatre?' by Saths Cooper

The Black artist (I shall refer only to Black artists and Black people, for whites are not my concern; my people, all Black people, are my sole involvement) is challenged by a mammoth task in racist South Africa. That of recognizing the problems facing Black people in this country, aiming at a solution to these problems and executing this aesthetically through the medium of his art.

Black theatre does not speak to the oppressors – it has nothing to say to them. Everything it has to say and do is for the ears and eyes of Black people only.

Black theatre is a theatre of liberation! It is the forum for an expression that would otherwise be stymied and witch-hunted out of existence. This expression is as honest and pure as a new-born baby. And the ogre that would take a sadistic delight in scuttling it into oblivion is the social system that prevails.

Black theatre is a challenge of confrontation! Confrontation with life in a racist society, and overcoming this mere meaningless day to day existence, this meandering in a state of flux, without goal and without fruition. Confrontation with the racist, and coming to meaningful terms with Black people in their idiom. Facing issues squarely in the face and accepting responsibility for the consequences. Black theatre is the theatre that is relevant to the needs of the people. But it will shock! Blacks who have been Rip van Winkles and Sleeping Sallies will be jolted back to reality. Blacks who have so far adopted a 'don't care' attitude will be forced to commit themselves to doing something about themselves and their surroundings.

By the foregoing it can be seen that Black theatre is made up of two levels: thought and action. The thought attempting to break down the psychological oppression that is presently shackling Black people, and the action defining Black people meaningfully, with all the physical and other attributes possessed by all human beings. The end result is the beautiful if startling whole that is Black theatre. Just as 'God created man in His own image' the Black artist must mould theatre in his own image, to *reflect accu-rately* his desires, his aspirations, his experiences, his life-style. And, in order to be always alive to Black people, Black theatre needs constantly to evaluate itself in terms of its achievements and its redundancies, so that its purpose is never lost, but remains firmly to serve the people.

We in this violently political country have a need for art. Not mere entertainment. But art which speaks to us in constructive terms about doing something about our situation NOW before it's too late!

And Black theatre is not an alternative to white racist oppressive art or a reaction to the system – it goes much deeper than that. It has to do with self-respect and asserting our dignity as human beings.

ITEM 2: **The Story of Bethal**

'The Story of Bethal'

In order to discover the truth about the way contracts are signed, Mr Drum himself decided to become a farm recruit. He was soon picked up outside the Pass Office by one of the touts or 'runners' who look out for unemployed Africans, and are paid for each man they collect for the agencies. He was taken to an employment agency, where he did not, of course, give his real name, Mr Drum, but adopted the name of George Magwaza. He said he had no pass, and, with many others, was told that he would be given a pass if he signed a contract to go and work out of Johannesburg: this is the normal way of dealing with people without passes. He chose to work on a farm in Springs, and was sent to __'s compound, where he waited nearly a day before he could sign the contract.

When the contract came to be signed the interpreter read out a small part of the contract to a number of recruits together, while the attesting officer held a pencil over the contract. No one asked the age of any of the recruits (they should have consent of parents if under eighteen), and Mr Drum was told nothing about whether his pay would be monthly or deferred, what food he was entitled to, or what length of shift he would work.

N A D African Clerk (calling roll of everyone on the contract sheet): You're going to work on a farm in the Middelburg district: you're on a six months' contract. You will be paid £3 a month, plus food and quarters. When you leave here you will be given an advance of 5s for pocket money, 10s.5d. for food, and 14s. 5d. for train fare. The total amount is £1 9s. 5d. and this amount will be deducted from your first month's wages. Have you got that?

Mr Drum and other recruits: Yes.

Clerk: You will now proceed to touch the pencil.

Mr Drum: But I was told before that I was going to be sent to a farm in Springs. Why am I now going to Middelburg?

Clerk: I'm telling you where you're going, according to your contract sheet, and nothing else.

So Mr Drum refused to touch the pencil when he reached the attesting officer, and was told to wait outside for his pass.

The other recruits then ran past the attesting officer, each holding the pencil for a moment, which was not even touching the paper.

Mr Drum, who can read and write English, had no opportunity either to sign his name or read the contract – but on his way out, he succeeded in obtaining a contract. As a result of holding a pencil for a second (50 recruits were attested in a few minutes), the recruits were considered to be bound to a contract. But in fact the contract had not been signed and had not been fully understood. So it seems that none of the contracts 'signed' in this way are valid at all (Native Labour Regulation Act of 1911, as amended 1949).

To find out what happened after the contracts had been signed, Mr Drum went to Bethal to obtain the facts at first hand. With a good deal of difficulty and sometimes at some risk, Mr Drum succeeded in talking to a large number of people most closely concerned with farm conditions, and carefully compared and checked the different accounts. Sometimes the farmers themselves were friendly, and at one farm Mr Drum was presented with a sack of potatoes. Mr Drum was very careful not to cause any trouble or enmity on the farms, and never tried to influence what people said.

Out of over 50 labourers interviewed on eight farms stretching from Witbank to Kinross, not a single labourer admitted that he was satisfied with the conditions. Those who did not express this view refused to comment altogether, for fear that they might be victimised.

Two-thirds of those consulted said they were sent to Bethal under false pretences:

they were either promised soft jobs in Johannesburg or on dairies in the Springs district, but they subsequently found themselves being made to alight at Bethal Station and told they were going to work there ...

Joseph __, who said he was 14 years old, told me he was recruited by X's agency, in the Northern Transvaal to work in a clothing factory at Springs. He was given an advance of 10s. for food and a train ticket, only to discover at Springs that he was going to work on a farm at Bethal at £2 a month.

Mzuzumi __ (30) says he was recruited in Natal by Siz'Abafane Employment Agency, given a 10s. advance and told he would work in Johannesburg. He had no pass then and accepted the six months' contract as a solution to his problem. Siz'Abafane's guide got their party to alight at Bethal Station in the dead of night and told them that is where they were going to work.

The pay on the farms is between £2 and £3 a month, and the food consists mainly of porridge, with meat sometimes once a week, if that.

Months are calculated on the basis of 30 full working shifts, excluding non-working days such as Sundays and public holidays, and the wages for the first month are spent in repaying train fares and money advanced to the labourers as a loan on recruitment.

For example, R __ F __ (60), employed on the farm of Mr B, was recruited by Z's Agency, of Johannesburg. He earns £3 a month and has a wife and four children to support at home.

His fare to Johannesburg was £1 6s. 11d., and his whole wages for the first month repaid this sum. He will have £15 3s. 1d. to his credit at the end of his six months' contract. But if he decides to return home he will be minus another £2 6s. 11d. when he reaches Louis Trichardt, which means that he will be left with £12 6s. 2d. in cash, or even less should he ask for tobacco or clothes on credit from his employer before that time, to say nothing of what he will spend on his journey home. And that for half a year's work ...

Very often the boss boys, themselves Africans, are tough and ruthless with the labourers, for if they are not they lose their jobs. Most boss boys are old employees who have acquired the important status of permanent squatters on the farms, where they live with their families and repay the farmer by working for him free of charge. They enjoy certain facilities, such as the use of a horse when they supervise the labourers.

At Bethal Hospital I found Casbert Tutje, from Cetani, Cape Province, who, together with three other friends, was also recruited by Siz'Abafane Agency in Natal.

From there they were sent to the farm of Mr X at Bethal, to work as labourers at £3 a month each. This happened in November, 1950.

The foursome was invited by a family squatting on the farm to a beer drinking party on Christmas Day, but, as the farmer would not grant them permission to attend, they left the compound without leave that morning. The boss boy had left the compound gate open.

The farmer sent for them and they were severely punished, then handed over the police. They were brought before the Court on a charge of desertion and sentenced to two months each. The farmer then arranged that their terms of imprisonment be served on his farm, where they were again thrashed by the boss boys severely.

It was in the course of this that Casbert sustained serious internal injuries which resulted in his being admitted to hospital from time to time. He gradually became weak and sickly, and spent more of his time lying in hospital than working on the farm. Before his contract expired in 1951 the doctor advised that he was unfit for farm work and should be sent home.

The farmer refused to pay Casbert his wages, however, stating that he had not completed his contract and still owed money for an overall purchased on credit; he would not give him his pass either. Casbert was given the sum of 10s. for food and told to leave the farm. And when he reported the matter to the Native Commissioner he was told that nothing could be done about it. His next alternative was to seek employment on another farm, with a view to obtaining a new pass and getting money to pay for his fare home.

Here, too, he had to sign another contract. He is not home yet. He has been a patient at Bethal Hospital since September, 1951, and is now diagnosed as a pulmonary TB case with little hope of recovery. He still has the new contract to complete ...

People living in Bethal tell of labourers who died of cold while deserting or simply while living in the compounds, and there are farmers who, probably because of their ruthlessness or otherwise, are known by such names as Mabulala (the Killer) and Fake Futheni (Hit Him in the Marrow), and so on.

Most of the compounds I saw look much like jails. They have high walls, they are dirty and are often so closely attached to a cattle kraal that the labourers breathe the same air as the cattle at night. Some labourers told Mr Drum that they are locked in at night ...

Farm prisons and contracted labour by-pass the normal need to attract men by improved working conditions and higher wages. They depend upon compulsion, not persuasion.

Most men who touch the pencil for a farm contract are hungry, ignorant and urgently in need of work. Once they have touched the pencil they have handed themselves over to an unknown area under largely unknown conditions.

It is obvious that care has been taken by the authorities to protect these people and equally plain that they have failed. For once men have 'signed' themselves away, itself often a trick, they are taken to an isolated farm where they are at the mercy of the farmer and his boss boys; brought back the police if they run away; and liable to every abuse with no chance of being able to protect themselves.

We ask, when farm life is so often satisfactory, what are the conditions which have given Bethal so fearful and exceptional a history – and we reply: it is the system, the farm contract system that has had so vile a result.

FROM: Mothobi Mutloatse (ed.), 1981. *Recon-struction*. Johannesburg: Ravan.

TIME TO DISCUSS

- ✪ How has each of these documents added to (or detracted from) your understanding of *Dark Voices Ring*?
- ✪ Had you heard of Black Consciousness before reading *Love, power and meaning*? If so, where or how?
- ✪ In your opinion, does *Dark Voices Ring* meet the requirements of Black theatre?
- ✪ Have you ever been at a political rally where poetry was read or a play performed? Describe the event and your feelings about it.
- ✪ Do you think that plays, poems or novels can motivate people to take political action?
- ✪ In your view, how should we evaluate *Dark Voices Ring*? Should we evaluate it in terms of its political message? Are its artistic and stylistic features relevant?
- ✪ If *Dark Voices Ring* failed to motivate its audience to react against oppression, do you think that it could be considered a failure? Explain.
- ✪ Do you think that writing which deals directly with social and political conditions has a responsibility to depict these conditions objectively and factually? Can you give a reason for your view?
- ✪ To what extent do you think Henry Nxumalo's 'The Story of Bethal' is a reliable historicial source?

Assumptions and Implications

It is difficult to think of a single major modern critical study that does not pay serious attention to context. Today, the connection between literary texts and their social, political, economic and cultural contexts is usually taken for granted. This is quite a change from the methods of criticism that were popular as recently as twenty years ago. It seems to represent something of a power shift in universities and may have emerged from popular demands that university syllabuses serve social interests. It also could be a response to the sheer fatigue of endless close readings – going beyond the text offers fertile opportunities for something new to be said.

Just as the performance we see on the stage is a particular production of a play text and just as a critical article on a novel represents a selective reproduction of that novel, so too is history a particular version of the past. As with any piece of writing, we need to ask from which perspective it is written. We need to be aware that it may be biased and that it may serve the vested interests of those people or groups which hold power.

As we have pointed out, history isn't neutral or objective. If you adopt a strictly Marxist approach, you need to be aware that this implies a particular world-view that has a specific agenda. In Chapter five, we discussed Marx's sense of modern society as a power struggle between the dominant capitalist class and the oppressed working class. An important aspect of this struggle is the use of ideology or 'false consciousness' as a means of maintaining capitalist power. Literature, in Marx's view, is one of the mediums through which ideology is transmitted. Later, in outlining Foucault's ideas about knowledge and politics, we noted how knowledge serves as a mechanism for social control. Foucault was not a Marxist, although he shares some ideas with Marxists. Nevertheless, the trend in contemporary socio-historical and socio-political approaches to literature has been to favour works that actively resist current systems of control. In the case of Marx this is clear. He argues that the proletariat must

USEFUL SOURCES ON CONTEXT

Aside from being aware that history is open to distortion, we need to ensure that our recovery of context is sufficiently detailed. Valuable sources on context include:

✪ Other literary texts from the period.

✪ Critical studies dealing with the period or subject. In the case of South African theatre there are at least three major published studies available: Robert Kavanagh, 1985. *Theatre and Cultural Struggle in South Africa.* London: Zed Books; Martin Orkin, 1991. *Drama and the South African State.* Johannesburg: Witwatersrand Press; and Loren Kruger, 1999. *The Drama of South Africa.* London: Routledge.

✪ Historical accounts (these are usually found between DEWEY REF NUMBERS 900–999 in a library catalogue).

✪ Reports of Organizations such as Amnesty International, the United Nations and, in South Africa, the Institute of Race Relations. Either search for these in a library or write to the organization concerned.

✪ Newspapers. Many libraries and the newspapers themselves have collections available on microfilm. In some instances, these date back several hundred years.

✪ Autobiographies and biographies of ordinary people and leading figures in politics, culture and society.

✪ Economic and statistical reports.

✪ Archives and History Museums. The State Archive in Pretoria and its branches in other major cities are useful. The National English Literary Museum in Grahamstown also has holdings on many Southern African writers.

✪ Oral testimony. When researching more recent works, you can interview people who were alive at the time or, better still, were involved politically or socially in the conditions you are investigating.

overthrow the bourgeoisie in order to bring about a classless society. Accordingly, Marxist literary critics favour works that advance these ideas. Foucault also sees analysis that actively challenges patterns of domination as 'valid'. Both attitudes suggest the existence of evaluative criteria from within the broad frame of contextual approaches to literature. Although Marxist critics claim that their method of analysis and evaluative procedures are objective – scientifically disinterested – this argument is not easy to sustain. Works that do not urge class struggle or expose the ideological underpinnings of social phenomena are often dismissed as hopelessly bourgeois. Clearly this is more than slapping on a bit of history to make a piece of criticism seem comprehensive. It involves a certain commitment to political change and a specific way of reading texts.

✪ What role do you think literature should play in political life?

✪ Should we expect literary works to make political statements?

✪ By emphasizing context, is there a danger that we might forget the artistic elements of literary works?

✪ Are the answers to questions such as 'what does this work mean?' or 'how does this text work?' to be found in contextual matters?

Views on literature and politics

MY VIEWS

I believe that literary texts should promote political change. I think that it is a writer's duty and a critic's responsibility to campaign for freedom – whether this is freedom from political oppression, or from stifling gender roles, or from discrimination on the basis of sexual preference or religious belief. I think that the vast global discrepancies between huge wealth and grinding poverty are obscene: they should be addressed. Literature needs to play an active role in doing this. What do you think? Should we expect writers to 'campaign for freedom'?

● SHOULD WE VALUE a work purely because it expresses the 'correct' political opinion?

A DIFFERENT VIEW

I believe all men have an obligation toward the political problems of their society, and I see no reason why writers should be excused from this obligation. What seems to me wrong – not wrong, but rather useless or stupid – is that those people for whom political action is primordial, is their true vocation, seek to exercise it through literature. Because literature is not a political instrument, or better, it is the worst political instrument. In such cases, it is preferable to tackle politics directly and not to waste one's time and energies making literature.

Mario Vargas Llosa

(FROM: 'An interview with Carlo Meneses' in Anne Freemantle (ed.), 1977. *Latin American Literature Today*, 329–330. New York: New American Library.)

MARIO VARGAS LLOSA'S international stature extends beyond literature into the realm of politics. He is the author of over a dozen novels and is often mentioned as a candidate for the Nobel Prize for Literature. His work is grounded in Peruvian culture but it also addresses problems that are relevant to other contexts. He is an outspoken cultural critic and political figure, who very nearly won election for the presidency of Peru in 1990. In 1998, he won the National Book Critics Circle Award in the United States for his last book of essays, *Making Waves*.

FOUR

PLAYS ARE WRITTEN to be performed. The play text is merely a blueprint for a stage performance. Poems and novels might be read aloud, but they are not primarily meant to be visual spectacles. Unlike other literary genres, plays are deliberately written so that they can be enacted on a stage. This presents several challenges to the literary critic. At the outset, we need to decide what our object of study is. Do we focus on the text of the play, which is not the finished product, or do we concentrate on the stage spectacle? If we opt for the text, we run the risk of ignoring an integral aspect of the creative process – the transformation of words into action. However, if we insist on looking at the play we have to deal with the fact that performance is ephemeral; it can change from night to night and from production to production.

The difficulty that we face in reading plays in performance is not too far removed from the issues that we face in criticism generally. All communication is open to interpretation and evaluation. Just as one production of a play may differ from another, so too do readings and evaluations of poems and novels vary. Similarly, a director's reproduction of the play text represents a specific interpretation of it. The director and actors do not merely restate the text objectively, they bring their own ideas to bear on it. Through voice, costume, make-up, movement and all the other devices of theatre, they give the play a particular feel or slant. On the basis of theatrical decisions, the role of a character may be foregrounded or turned from serving a comic function to making a more serious statement. While we may like one production of a play more than another, it isn't easy to say which is the correct or only permissible version. In the same way, reader-response theorists have supported the idea that each literary text should produce a range of possible readings. Who is to say why a reading is right or wrong?

This complicates the situation for students of drama, though, especially if they consider the play in performance. This is because the 'thing' – the play on stage – that they are studying is a particular production of a text that is already an imaginative reproduction of society or life. Whatever the theatre critic says is an interpretation (the critic's view) of an interpretation (the director and actor's view) of an interpretation (the playwright's view) of 'reality'!

REALITY ➤ PLAYWRIGHT ➤ DIRECTORS AND ACTORS ➤ CRITIC

Here, it's important to distinguish between two types of evaluation. The first type is based on the skill of the actors, which may have nothing to do with the quality of the play. A bad bunch of actors can turn even the most promising text into a failure. This does not mean that the play itself is weak. In the hands of skilled professionals, the same text may result in a production that captivates and delights its audience. The second type of evaluation deals more with the dramatic potential of a particular piece. This evaluation is based on seeing the possibilities that a text provides for successful production. It looks at whether, under the right circumstances, the written word can be successfully translated into a powerful (and entertaining) theatrical occasion. As literary critics, we are concerned with the second option. How, then, do we go about evaluating a play in performance?

The main elements that we need to consider when discussing performance are the setting (particularly the use of stage décor), the appearance of the actors (what they wear, their make-up, how and when they move) and the manner in which the lines are delivered (intonation, pace, volume and so on). Each element of performance contains theatrical 'signs'. The study of signs, in the theatre and in literature generally, is called semiotics. As they advance in their studies, students of the theatre need to master the semiotics of performance. They need to learn what various theatrical signs mean, and how they can be deployed to achieve certain effects. We are not going to discuss the jargon of theatrical semiotics. It is enough, for the present, to keep in mind the fact that all the elements of theatre potentially operate on a symbolic level. Thus, a certain set design or a specific way of walking may convey meanings that the audience recognizes without anything having to be said. Lighting, costume and gesture add to the effect. By carefully manipulating theatrical elements, such as gesture or lighting, it is possible to create a tension between what a character says and the actual meaning that is conveyed. Thus, characters may claim to be honest or sad, to be friendly or angry, while their gestures or actions suggest that they are not what they claim to be. This discrepancy or gap between what is said and what is portrayed on stage is sometimes referred to as dramatic or visual irony. The speech from *Julius Caesar* offers the actor who is playing Antony an ideal opportunity to behave in a way that emphasizes the irony of his words. Moreover, by subtle changes in lighting or by placing the dead body of Caesar in a particular position, the production may draw greater attention to the magnitude of the deed that Brutus has carried out. When reading a text, we need to imagine the ways in which it might be transformed on stage. In *Dark Voices Ring*, for instance, there is ample opportunity for the actors to reinforce the meaning of their words through mime, song and movement. Moreover, as the action shifts from the present time in the hut to the past incidents on the potato farm, so would the use of lighting and sound change.

Apart from visualizing the play in performance, critics often consult records of various productions. Where a play has been videotaped or filmed for television, we can examine specific productions in close detail. If the actors and the director are available, they can be interviewed. In many instances, it is also possible to track down set and costume designs, lighting plans or at least to find photographs of the staged work. Usually, though, our information on the performances of a play comes mainly from newspaper reviews or other written records. Although these pose their own limitations (they are by nature subjective and reflect the values and interests of the reviewer), they can offer valuable clues into both individual interpretations of a play and the audience's response. Thus, we can gauge how plays have spoken to audiences in times and contexts other than our own.

Let's look at Christopher Prophet's review of *Dark Voices Ring*, which was performed as part of a double-bill with another Mda play, *We Shall Sing for the Fatherland* at the People's Space Theatre.

RAW LOCAL PLAY WITH MESSAGE

AT THE PEOPLE'S SPACE: *Dark Voices Ring* (with Nomhle Nkonyeni, Nkos'omzi Ngcuknana and Sam Phillips: directed by Rob Amato and Nkonyeni) ... Both plays by Zanemvula Mda.

This is raw, indigenous theatre with a rough and vital message.

Dark Voices Ring centres on three characters caught in an emasculating farm baas/labourer situation. The Old Man, once the boss boy on Baas van Wyk's potato farm, sits silent, eyes closed, wrapped in a blanket, on a tatty grass mat. His wife, by his side comforts him.

The son-in-law, who never got to marry her daughter, bursts in upon the dream her daughter still lives. He has come to say goodbye. He is going to live his dream, to fight for freedom on the northern borders.

Before he goes he wants her to come back to reality, to face the world and fend for herself.

She is forced to face the truth. When prison labour was introduced her boss boy husband drove the prisoners hard. He enjoyed it and made the white jailers laugh.

He drove them too hard. They began to sing: 'What have we done, black people? Why do they whip us like dogs?' The prisoners rise up, beat up the boss boy and burn his hut. Inside is the wife's baby girl, her only daughter.

The son-in-law will fight for freedom and redeem their

honour. 'Go, my son,' says the woman. The Old Man's eyes open and he smiles.

The play, short and powerful, is carried by Nomhle Nkonyeni's perfomance as the old grief-stricken woman ... Nkos'omzi Ngcuknana as the son-in-law is stiff and somewhat forced. Sam Phillips, the Old Man, sits very still.

Mda's play, of course, is the thing. It is a cry from the heart and as such I felt it deserved more.

✪ What criteria has the critic used to evaluate the play?
✪ Think about the critic's use of words such as 'raw', 'rough', 'indigenous' and 'vital'? What do they imply about the play? How might the review have differed if it had been performed by a different cast?
✪ Write a review of the text of *Dark Voices Ring* for your local newspaper. (Reread Chapter two to help you.) Then consider how your review differs from the review printed above.

⬤ Assumptions and Implications

By emphasizing the play's production and reception we are moving still further away from the idea that texts have a single and inherent meaning and value. In its place we are opening up an opportunity for multiple interpretations and widely varying evaluations. Instead of claiming that meaning and value are embedded in the text, we are saying that they are functions of the reader, director or audience. Potentially, there are as many interpretations as there are readers or, in other terms, as many productions as producers. (You may want to look now at what we wrote about reader-response criticism in Chapter one.)

In practice, this is not the critical free-for-all that it suggests. Generally, there are limits that 'sense' places on the meanings of a text, if not on its value. No one in their right mind would claim that *Dark Voices Ring* is about something as removed as rose-picking in China! Moreover, our readings of texts are shaped, in part (perhaps entirely), by the values and assumptions that we share as individuals and as societies. The mere fact that we are raised in families and communities suggests that ideas and the meanings that we give to people, actions and things belong to a shared code of perceiving and understanding. As Foucault has shown, the influence of Western social institutions, including the family, is to establish clear, shared norms. It is likely then that all members of a predominantly Christian society will see a cross as representing the Church and Christ's crucifixion, or a dove as a symbol of peace. Of course, other cultural groups may use these symbols to signify different meanings. Nevertheless, within a specific community there is likely to be broad agreement about what things mean. In effect, this limits the possibilities for a play's production and interpretation. Playwrights, directors and actors tend to draw on symbols and actions that are culturally familiar. It is not much use using an elaborate symbol or gesture if no one knows what it means. Similarly, critical claims that appear to bear no relation to the evidence are unlikely to be taken seriously.

Of course, playwrights also undermine our shared under-standings of what signs mean. For instance, we briefly considered an episode from *Gulliver's Tales* in the previous chapter. Here, Swift takes categories that we know – horses and humans – and gives them characteristics that break down our usual connections and assumptions. He violates our shared understanding of how each should behave and thus draws attention to his real target – human assumptions of civilized rationality. Playwrights often do they same thing by drawing on conventional icons and symbols, which they then imbue with new resonances of meaning. Perhaps, this is what makes literature more enjoyable and entertaining than a scientific description of the nucleus of an atom.

● DO YOU THINK that if enough people think a play is good, it must be?

FIVE

ALL LITERATURE IS political. Even works that do not mention politics directly are political. Jane Austen's *Pride and Prejudice* (1813), which deals with the lives of the five Bennet sisters and their search for suitable husbands, has the same potential to be interpreted politically as Mda's *Dark Voices Ring*, which culminates in a call for young back men to join the armed struggle for liberation. Any work that deals with the position of men and women in society, however obliquely, makes a political statement. Even silence can be interpreted politically, for if we fail to speak out against injustice our silence can be construed as tacit support for it.

Clearly, *Dark Voices Ring* belongs to a category of novels, poems and plays in which overt political statements are made. Political and polemical writing is not unique to South Africa or to this time in history. Writers all over the globe, on many diverse occasions and in widely ranging circumstances, have found it necessary to speak out against authority, oppression and injustice. This, as you know, is often a dangerous stance to take. Regimes, especially of the authoritarian or dictatorial kind, have never been afraid to act decisively to stop criticism, especially if they can represent it as dangerous propaganda. If you dare to make political statements you generally can expect a political reaction.

In 1980, the Ravan Press collection in which *Dark Voices Ring* was published received a bad review of a very unwelcome sort. This review had little to do with the play's artistic merits or its literary quality: it was based solely on its political content. The first publication of *We Shall Sing for the Fatherland and Other Plays* was banned in South Africa under Section 47(2)(e) of the Publications Act of 1974. This section of the Act empowered the (Nationalist) government to restrict any publication that, in its view, was 'prejudicial to the safety of the State, the general welfare or the peace and good will'. The offending play was *Dark Voices Ring*. The problem for the censors who made this decision was the Man's rallying call for an armed struggle, which was perceived as posing a threat to a state that had shown little restraint in its treatment of dissidents. Later, the banning was overturned by a Publications Appeal Board and the play was released for publication, but their findings concluded ominously by noting that 'any public entertainment which could result from the present plays is not a subject of this adjudication'. Clearly, the authorities wanted to keep their options open in case the play inspired a rush for the border posts! Perhaps, too, they realized that actions speak louder than words.

I have included an uncorrected copy of the Report of the Publications Appeal Board finding on *We Shall Sing to the Fatherland and Other Plays*. The reports of the Committee of Experts, in Afrikaans, were appended to the original document, but I have not reproduced them. I would like you to read through this report carefully, answering the questions in the margin as you go.

IN THE PUBLICATIONS APPEAL BOARD

Date of meeting: 1981-09-30   Case No: 53/81

DIRECTORATE OF PUBLICATION   Appellant

v

THE COMMITTEE OF PUBLICATIONS   Respondent

In regard to the publication: "WE SHALL SING FOR THE FATHERLAND AND OTHER PLAYS"

Appeal Board: Prof dr JCW van Rooyen (Chairman)
                 Genl J Joubert
                 Mr CD Fuchs
                 Dr CE Pretorius

SUMMARY

Plays not undesirable. Publication to be distinguished from public entertainment.

The publication "We shall sing for the Fatherland and other Plays" by Zakes Nda [sic] (Ravan playscripts 6) was found undesirable by a Publications Committee within the meaning of sect. 47(2)(e) of the Publications Act 1974 as amended. The Directorate of Publications appealed to the Publications Appeal Board against this decision, requesting the Chairman to appoint a Committee of literary experts in accordance with section 35(A) of the Act. The positive report of the Committee of Literary Experts is appended to this report.

The interpretation of sect. 47(2)(e) of the Publications Act is dealt with in decisions such as <u>Staffrider</u> (122/80) and <u>Forced Landing</u> (45/80) and need not be repeated here.

Only the second of the three one-act plays in this volume, <u>Dark Voices Ring</u>, was found to be undesirable by a Publications Committee in terms of sect. 47(2) (e) of the Publications Act. It was their opinion that the play is prejudicial to the safety of the State, peace and general public welfare in so far as it constitutes, in effect, a rallying call for black men to go north to join "those men in the bush fighting with their guns and dying to save us" and certain passages in it "constitute incitement to join 'freedom fighters' (terrorists?) against RSA".

On the other hand, the Committee of Experts appointed by the Chairman of the Appeal Board in terms of sect. 35 (A) of the Act could find nothing in the play at issue that could incite or inflame the feelings of Blacks so much as to justify the need for it to be prohibited.

In that there may be possible incitement in <u>Dark Voices Ring</u>, its theme (or message) should be noted at once. Dreams may well be regarded as the theme: on the one hand, those of an old black woman, who, as an outcast, with her old, mute, and apparently mentally dead husband, lives

[sic]: this error was printed in the original text

● DO YOU THINK the State has the right to ban works on these grounds?

● DO YOU AGREE that the theme of the play is 'Dreams'? Why?

only with her memories, and, on the other, those of a young man, her son in law, whose main concern is the future and the freedom of his people. As the play is mainly concerned with the wrongs of the past, there is little room in it for elaborating on what the young man thinks should be done in future.

● IS THIS a weakness in *Dark Voices Ring*?

The old man, Kaptein, was the boss-boy on the farm of a certain Jan van Wyk. He was scorned and rejected by other blacks because of his trying to ingratiate himself with the farmer through his severe discipline. He was especially hard on black prisoners, whom the farmer had employed to dig potatoes, "with their bare hands", so that the work could proceed more quickly and his crops be increased faster. The prisoners rose, burnt the boss-boy's hut down, his daughter Nontobeko, to cinders, and the farmer's house as well. Though the three leaders were executed, Kaptein himself, ever after that, was mute and mentally withdrawn. The old woman, however, apparently managed to survive in a world of illusions about her husband's former importance and their comparative well-being through his loyal service to his master. These illusions, the young man, about to go north to join the fight for freedom, has come to destroy. There are, inevitably, derogatory references in the play: to the farmer's greed in wanting to make profits fast and, therefore, egging his boss-boy on to use his whip whenever the prisoners sneered at him, to the warders' "drinking and enjoying themselves on the stoep, because they had some-one they could trust, someone who could do what even they couldn't do", and to the boss-boy's being like "the black officials of this regime", who "are doing their duty, whilst they mow down peaceful children marching down the street …"

● WHY WAS THE Appeal Board comfortable with 'the language of protest' in the written text of the play?

This is the language of protest, with which the Appeal Board has become familiar in black and other political writings, for instance in such publications as Forced Landing (45/80) and Staffrider (122/80), and which the

• DO YOU think
the appeal board
would have ruled
differently if the
whites had played
a more prominent
role in the play?

• WHAT IS
YOUR opinion of
this view of the
play's impact?

Board has found, in those instances, to be "a typical feature of the South African political scene". It does not constitute direct political incitement and is not likely to harm relationships between sections of the populations because much of the disparagement is directed at the old black man, who, as one of the three characters in the play, is really the villain of the piece. The white farmer and the warders, who are merely referred to in the dialogue, have no direct part in the plot.

The references to "the bush war in the north", a "war of freedom", and to "those who are dying in order to us" are too vague to have any effect as "a call to arms". Nowhere is there an indication of where the play is set, and, consequently, both "the bush" and "the north" are unlocalized. Nor are those "who are dying" at all specified as representatives of a national or tribal or political group. The sentiments expressed by the young man cannot, therefore, be associated specifically with "terrorists" or "freedom fighters" that have been declared enemies of the State.

Whatever political impact the play may have been intended to have is considerably lessened both by the form of the play and by the treatment of the plot. The Committee of Experts quite rightly remarked upon the weakness of the ending: "'n hinderlike klemverskuiwing in die karaktervoorstelling van die vrou" [a troublesome change of emphasis in the characterization of the woman], who, in contrast to her having been an individual with her own distinct point of view, becomes merely the mouthpiece of the author. The old man, in the end, is rendered equally unconvincing. Portrayed throughout as an object of contempt, appropriately punished for his inhumanity, by apparently losing all awareness and the power of speech, he smiles broadly with his eyes brightening up and his face full of life, when he hears the young man say that he is going to fight for both the old woman and himself. The transformation of the two old people is too sudden to be true. It cannot win any sympathy for them or more sympathy for the young man. There is, in fact, too little character development in the play to enable what the young man symbolises or represents to become compelling or convincing. He is no more forceful at the end of the play than he was at the beginning.

The one-act form of the play contributes to its ineffectiveness as a medium of propaganda or as "a rallying call", if that is what it was intended to be. There is no room in the play for the clear development of ideas to illustrate a line of thought to be adopted or a course of action to be taken. As the Committee of Experts has rightly remarked, the cause of and the need for the conflict with which the play is concerned hangs, vague and academic, in the air and is nowhere presented or developed as a dramatic actuality. The play, because of its brevity, is too slight to convey an enduring message to its audiences, certainly not a message undesirable in terms of sect. 47(2)(e) of the Publications Act.

In conclusion it must be pointed out that any public entertainment which could result from the present plays is not the subject of this adjudication. The present appeal only concerns the play as a publications with its own particular likely readership, which is a limited and intellectual one.

The appeal succeeds and the decision of the Publications Committee that the publication is undesirable, is set aside. Genl Joubert voted in the minority.

NJG SABBAGHA and

[signed]
JCW VAN ROOYEN
CHAIRMAN: PUBLICATIONS APPEAL BOARD
1981-09-05

• HOW DOES THE Appeal Board's view of the play differ from Andrew Horn's?

• WHAT EVALUATIVE CRITERIA has the Appeal Board used in judging the play? Do you agree with these criteria?

In my opinion, the Appeal Board were very poor literary critics, which is fortunate for the play because they unbanned it and it passed into general circulation. I think that *Dark Voices Ring* constitutes an important and powerful political statement, especially at a time of tremendous oppression. Moreover, the play satisfies my preference for literature that challenges all forms of discrimination. I value the play's political perspective. I also happen to think that it is a well-crafted work and that it conveys its message with economy and power. But my enthusiasm for the play shouldn't stop you asking questions both about the play or my sense of literary value.

✪ What value (if any) do you think the play's political message has in the context of the 'new' South Africa?

 A personal reflection

When I praise *Dark Voices Ring*, I wonder whether my own political ideas have clouded my interpretive and evaluative capacity. Have I allowed my political beliefs to influence me to such an extent that I can no longer see the subject, except in terms of these views? In other words, perhaps I value this play because its political position matches my own political beliefs. If this is so, I wonder whether I am prepared to tolerate political positions – in literature and in literary criticism – that are opposed to my own. Most painfully, I wonder what I am saying about myself when I express these ideas. Am I saying, 'Look at me. I'm a broad-minded, progressive liberal and I believe in human rights!'? Am I saying, 'I oppose oppression and always did'? Is all of this an exercise in 'political correctness' – a statement of my credentials in the 'new' South Africa? I cannot deny that these things are part of the sub-text of my stance. Consciously, by valuing *Dark Voices Ring*, I am separating myself from authoritarian versions of political expression. Then I wonder why I didn't join those people, black and white, who went north to fight for liberation. What do I do now to protect the weak, uplift the poor or free the oppressed? And I wonder if my commitment to freedom, justice and equality is a veneer – as thin and insubstantial as the paper on which this is written.

- What does your *attitude* to literary and cultural expression say about *you*?
- What obligation do you feel towards actively promoting social justice and equality?

. . . .

EVERY FEW YEARS or so, the department I work in goes to war. We don't actually kill anyone, although we might feel like it. Nor do we don army uniforms and carry guns. Instead we engage in a high-intensity, low-key combat over what we are going to teach and when; who's going to teach what and how; which books and ideas are worth studying and which should be trashed. Often the lines are drawn around English for Academic Purposes and English for Special Purposes, but even within these groups there are minor skirmishes. Sometimes, language study is pitted against literary study, which in turn vies for prominence over cultural study. All in all, it can be a very bruising experience, because these battles can mean the difference between getting a promotion or missing one, getting funding or losing out, being overloaded with work or having an easy time at the university's expense. In essence, these are political battles. They have very little to do with whether we should teach Milton in place of Donne or whether our students need to be taught the fundamentals of the business report instead of the argumentative essay. Rather, they are closely linked to the position and status that individuals enjoy in the department and in the larger academic community. If you are lucky enough to teach a popular course or to get your module prescribed as a part of the core studies of a programme, your position, for the meantime, is secure and your access to adequate staffing and resources is assured.

The literary critical battles in academic journals are similar. Success is when your colleagues and peers admire your intellectual depth and your ability to spot and demolish a spurious argument blindfolded at five hundred metres. Some of the

SIX

THE POLITICS
OF LITERARY
STUDY

debates that academics engage in have little to do with 'advancing knowledge' or 'truth'. Rather, they are elaborate displays of intellectual prowess. It is notable that a university, which should have as its purpose the preservation and extension of knowledge, does little to ensure that its deliberations are intelligible to the wider public. Academic style is often difficult and riddled with jargon. Instead of making things clear, it ties them up in maze of convention and overblown expression. I am not excusing myself from these practices. I am as guilty as the next academic. We all are. Even though we have written this book to be as engaging and accessible as possible, I am sure that at times we've fallen into the trap of trying to impress rather than to explain. I know that we're all conscious, as authors, of our colleagues' judgements and the criticisms we might bring on ourselves. By this, I don't want to deny that we have dealt with many difficult concepts and that the types of relationship we've explored are complex. Nor do I want to suggest that one does not have to pay a high price in effort and sacrifice to gain an advanced education.

But is it worthwhile? As I write the closing words to this chapter, twenty-one hostages are being held by separatist guerrillas in the Philippines. Eritrea and Ethiopia are at war, while millions of nomadic people face starvation in the latter country. Sierra Leone is on the brink of another civil war – the last one resulted in horrific slaughter and maiming. The Democratic Republic of Congo is in ruins and Zimbabwe faces a land crisis. Closer to home, crime and violence tax the efforts of the police and millions of people live without adequate shelter and enough food or work. Even more frighteningly, the Aids pandemic threatens to wipe out a quarter of the economically active population, even as it kills thousands of children. The Western superpowers seem to increase in wealth and technology, while the underdeveloped countries of Africa, South America and parts of Asia continue in poverty. Global warming, the devastation of the rain-forests, the extinction of the great apes ... I could go on. The world seems to be facing crises (and opportunities) of an unprecedented nature. In the face of these realities, how can we justify the study of literature and culture? Why bother ?

We are concerned with literature and culture because we believe that knowledge in these fields is uplifting and empowering. The arts offer us, individually and collectively, a chance to transform ourselves and to aspire to new heights of thinking and being. They have the capacity to change our lives. Through the imagination and through the critical eye of the writer and the scholar, we are able to perceive old things in new ways and new things in the light of the past. More importantly, literature and the study of culture pose a challenge to convention and tradition. For all its stolidness, in the long run literature has little tolerance for habit or polite conformity. Rather, at its best, it encourages us to scrutinize ourselves and our beliefs and constantly to re-invent ourselves in response to an ever-changing environment. Knowledge is power and knowing that knowledge is power can make us more powerful. I am not thinking here of power in terms of wealth or repressive control, but power as a liberating force from the various types of mental and physical oppression that threaten our happiness and security. I am particularly thinking of the power of the humane conscience that, despite the deterministic philosophies of many social theorists, can recognize 'the evil that men do' and react against it. It is to open the doors to this power of conscience and consciousness that we have written this book.

References and further reading

Adey, David, Beeton, Ridley, Chapman, Michael and Pereira, Ernest, 1986. *A Companion to South African English Literature.* Johannesburg: A.D. Donker.

Berger, John, 1972. *Ways of Seeing.* Harmondsworth: Penguin.

Bloom, H. and Trilling, L. (eds.), 1973. *Romantic Poetry and Prose.* London: Oxford University Press.

Chapman, Michael (ed.), 1986. *The Paper Book of South African English Poetry.* Johannesburg: A.D. Donker.

Coetzee, J.M., 1999. *Disgrace.* London: Secker & Warburg.

Davidson, Margaret, 1986. *I Have a Dream: The Story of Martin Luther King.* New York: Scholastic Inc.

Eagleton, Terry, 1986. *Against the Grain: Selected Essays.* London: Verso.

Eco, Umberto, 1986. *Travels in Hyperreality*, trans. William Weaver. London: Picador.

Freemantle, Anne (ed.), 1977. *Latin American Literature Today.* New York: New American Library.

Foster, Leonard, 1969. *The Icy Fire.* Cambridge: Cambridge University Press.

Garrow, David J., 1981. *The FBI and Martin Luther King, Jr.* Harmondsworth: Penguin.

Foucault, Michel, 1991. *Discipline and Punish: The Birth of the Prison.* Harmondworth: Penguin.

Herbert, L. Dreyfus and Rabinow, Paul, 1982. 'The Subject and Power' in *Michel Foucault: Beyond Structuralism and Hermeneutics.* Brighton: Harvester Press.
Hobbes, T., 1987. *Leviathan.* London: Dent.

Kavanagh, Robert, 1985. *Theatre and Cultural Struggle in South Africa.* London: Zed Books.

Kruger, Loren, 1999. *The Drama of South Africa.* London: Routledge.

Lakoff, Robin, 1990. *Talking Power*. New York: Basic Books.

Lehman, David, 1991. *Signs of the Times: Deconstruction and the Fall of Paul de Man*. London: André Deutsch.

Madoc-Jones, Beryl and Coates, Jennifer (eds.), 1996. *An Introduction to Women's Studies*. Oxford: Blackwell.

Mda, Zakes, 1990. *The Plays of Zakes Mda*. Johannesburg: Ravan Press.

Miller, James, 1994. *The Passion of Michel Foucault*. London: HarperCollins.

Moog, Carol, 1990. 'Are they selling her lips' in *Advertising and Identity*. New York: William Morrow.

Oates, Stephen B, 1985. *Let the trumpet sound: The life of Martin Luther King, Jr*. New York: The New American Library.

Orkin, Martin, 1991. *Drama and the South African State*. Johannesburg: Witwatersrand Press.

Orwell, George, 1979. *Nineteen Eighty- four*. Harmondsworth: Penguin

Schreiner, Olive, 1985. *The Story of an African Farm*. Johannesburg: A.D. Donker.

Schulke, Flip, 1995. *He had a Dream: Martin Luther King, Jr., and the civil rights movement*. New York: Norton.

Swift J., 1975. *Gulliver's Travels*. Harmondsworth: Penguin.

Thiong'o, Ngũgĩ wa, 1981. 'Literature and Society' in *Writers in Politics*. London: Heinemann.

Copyright acknowledgements

The publishers and authors would like to thank the following individuals and organizations for use of photographic and textual material:

Sue Clark for 'Biting Through', first published in *The False Bay Cycle*, 1996, Cape Town: Snailpress, copyright © Sue Clark, 1996; Mongane Wally Serote for 'For Don M. – Banned', first published in *Tsetlo*, 1974; Michael Chapman for 'The Chameleon Dance' published in *The Paper Book of South African English Poetry*, Johannesburg: A.D. Donker; Vanessa Cowling for photograph of Sue Clark; The Unisa Art Gallery for photographs of *The Artist's Mother* by Gerard Sekoto, *Louis Maqubela* by Durant Sihlali and *Double Woman* by Walter Battiss; U.A. Fanthorpe for 'Patience Strong'; The Plomer collection of Durham University Library for photograph of Olive Schreiner; *SL* Magazine for Clique advertisement; Ebony Sales for Opium advertisement; John McIldowie and Associates for QC Classic Footwear advertisement; Ken Payne Opticals for Sting Occhiali advertisement; A&M Records, Inc. for 'Sister Moon' by Sting, copyright © 1987; Christine Marshall for 'Imagination/Verbeelding'; Grace Nichols for 'Like a Flame'; Fleur Adcock for 'Against Coupling'; Joan Larkin for '"Vagina" Sonnet', first published in *Housework*, 1975, New York: Out & Out Books, copyright © Joan Larkin, 1975; Lisel Mueller for 'A Nude by Edward Hopper' published in *Rising Tides*, New York: Washington Square Press, c/o Simon & Schluster; Alta for 'i would write a love story', copyright © Alta c/o Shameless Hussy Press, Berkeley California, 1971, 1983; Marge Piercy for 'Rape Poem', published in *Circles in the Water*, 1985, New York: Alfred A. Knopf, copyright © Marge Piercy, 1985; Terry Eagleton for 'The Ballad of English literature', published in *Against the Grain: Selected Essays*, 1986, London: Verso, copyright © Terry Eagleton, 1986; Zakes Mda for *Dark Voices Ring*, published in *We Shall Sing for the Fatherland and Other Plays*, 1980, Johannesburg: Ravan Press, copyright © Zakes Mda, 1980. Remaining sources are cited in the text.

Every effort has been made to trace copyright holders. Should any infringements have occurred, the publishers would be grateful for information that will enable them to correct these in the event of a reprint.

Index

close reading, 175–7, 300–3
clues, 12
codes of meaning, 102, 117, 138
codes of significance, 102
Coetzee, J.M. *Disgrace*, 55, 57–9
    critics' evaluations, 59–63
cogency, 12, 16–18
coherence, 12–13, 17–18
communal social structure, 203
communication, 2–6
    in media, 99, 106
    in poetry, 9–10
communications model, 5, 97
complex language, 251–2
composition technique, 64–5
concept, 109
condoms and advertising
        techniques, 135–6
conflict limitation, 214
connotation, 11, 115–16, 151–2,
        170
context, 4, 6, 22–3, 33, 38, 42
    historical, 34–5
    sources, 320
contextual evaluation, 65–7
    historical, 312, 319–20
contradictions, 13
conventions of language, 5–6
courtly love, 154, 157–63
Coward, Rosalind, 134
criticism *see* literary criticism
culture and language, 249–50

D
de Man, Paul, 48
decision making, power of, 234–5
Declaration of the Rights of Man and of Citizens, 227
'defamiliarization', 10
demeanour, 129
democracy, 226
denotation, 151–2
Derrida, Jacques, 46
desire, 95, 120–1

and advertising, 112–13, 116,
        124–5
and poetry, 142, 146, 158–63,
        170–2, 177–8
disciplinary power, 261–4
discourse, 46, 96
    of desire, 145
    of love, 142
    South African, 65
Divine Right of Kings, 225
Donne, John, 74
Drayton, Michael *Since There's No Help*, 156

E
Eagleton, Terry, 39
    The Ballad of English Literature, 311
Eco, Umberto, 102, 128–30
education and language, 250–2
Eliot, T.S., 74, 311
emotive language, 239–40
Engels, Friedrich and Karl Marx
    The Communist Manifesto,
        229–30
English language, 248–9
enjambment, 11
enjoyment, 25
Enlightenment humanism, 260
eroticism, 142, 185
ethics in advertising, 121–2
evaluation, 37, 52–3
    biographical, 67–8
    contextual, 65–7
    feminist, 68–9
    formalist, 64–5
    media, 97–8
    'play in performance', 322–4

F
fact files
    Booker prize, 55–6
    Mda, Zakes, 303–9
    South Africa (1948-1980),

        313–18
family unit, 203–6
fashion, 128, 130
feminism, 45
    see also under evaluation;
        interpretation; literary
        criticism
Ferguson, Gus, 31–2
figurative language, 240
Fish, Stanley, 25, 42
flattery, 140, 147–8, 154, 167–8, 190
    power of, 239
foreign language
    interpretation equals transla-
        tion, 6–7
formalist evaluation, 64–5, 121,
        123–4, 299–302
Foster, Leonard *The Icy Fire*,
        167–8
Foucault, Michel, 46, 48, 95,
        253–68, 319–20
    Discipline and Punish, 258–9
    power analysis and
        literary/cultural critics,
        271
France *Declaration of the Rights of Man and of Citizens*
        (1789), 227
'free and equal', 227
Freud, Sigmund, 118–20
Fugard, Athol, 78

G
Gadamer, Hans-Georg, 24, 43
'gay' *see* homosexual
gender, 118, 129–30
    and language, 16
    in literary criticism, 15, 68–9
    and power relations, 234–5
generalization in speeches, 241–3
government's rights, 228
Graham, Winnie
    and *Disgrace*, by J.M. Coetzee,
        62–3

historical hermeneutics,
43–4
Marxist criticism, 42–3
New Criticism, 41
poststructuralism, 46–8
postcolonial criticism, 44–5
psychoanalytic criticism, 48
queer theory, 48
reader-response criticism, 42
Russian Formalists, 40 (*see*
also formalist evaluation)
literary theory, 36, 39–40
literature, 337
communication process, 37
social/personal function, 38
textual analysis, 287–302
Livingstone, Douglas *Giovanni*
Jacopo Meditates (*on an early*
European Navigator),
189–90
Llosa, Mario Vargas, 321
Locke, John, 219–20
lotus flower in love poetry, 151
love, image of, 96
courtly love, 154–7
in Greek myths, 141–2
Petrarchan tradition, 154
Platonic tradition, 152
love poetry
as persuasive discourse, 142–3,
146–50
as portrait of beloved, 140,
143–7, 150–1, 157–62, 165–8,
170–3
theme of love lost, 155–7
writing techniques, 181–3
see also poetry

M

Macherey, Pierre, 42
'madness'
Foucault's theories, 255–7
males
attitudes to adverts, 126

'male gaze' phenomenon,
130–3, 137
marketing and anxieties, 135
Marshall, Christine *Imagination/*
Verbeelding, 174–7
Marvell Andrew *To his Coy*
Mistress, 157–63
Marx, Karl and Friedrich Engels
The Communist Manifesto,
229–30
Marxism, 312
Marxist criticism, 42–3, 319–20
mass media *see* media
Mda, Zakes, 303–9
Dark Voices Ring, 272–86
Dark Voices Ring: banning of,
328–34
meaning, 4, 6, 37–8, 101–4
in poetry, 8
systems of, 130–2
see also codes of meaning
media, 95, 98, 106
message construction, 99, 105
techniques of literary criticism, 97
medium (definition), 100
metaphor, 11
military, 261
mimesis, 15
misinterpretation, 7
Mitchell, James *Gay Epiphany*,
193–4
monarchies, 225–6
Moog, Carol, 117–18
Morphet, Tony, 89
review of *These Things*
Happen, by Shaun de
Waal, 89–92
Morris, Michael
and *Disgrace*, by J.M. Coetzee,
62
Moscow Linguistic Circle, 10
Mueller, Lisel *A Nude by Edward*
Hopper for Margaret Caul, 188

Mutloatse, Mothobi *Forced*
Landing, 80–1

N

Nathan, Manfred *South African*
Literature, 77
Neanderthal people, 209–12
New Criticism 'school', 40–1,
75–6, 175, 299–302
Ngũgĩ wa Thiong'o *see*
Thiong'o, Ngũgĩ wa
Nichols, Grace *Like a flame*, 178

O

'offensive' writing, 184–5
'official' language, 248
onomatopoeia, 183
opinions, 32–3
oppression, 228
oral culture/orature, 44, 75, 77
Orwell, George *Nineteen Eighty-*
Four, 244–6
Ovid, 154

P

Paine, T.
Common Sense, 226
Declaration of the Rights of
Man and of Citizens, in
The Rights of Man, 226–7
Panopticon, 261–4
Paton, Alan, 78
patriarchal power, 130–2
penal reform, 260
perception, 10
performance poetry, 197, 199
Petrarch, 154
courtly love tradition, 155–7
Piercy, Marge *Rape Poem*,
199–201
plaasroman (farm novel), 65
Plato, 38
Platonic tradition, 152
plays

reviews, 324–6
theatrical performance
    evaluation, 322–4
plot analysis, 292–9
poetic devices, 11
poetry
    interpretations, 173
    meaning, 8
    reviewing, 32
    writing techniques, 100–1,
       181–3
    see also love poetry
'political liberty', 228
political speeches, 239–43
political systems
    proposed by J.J. Rousseau,
       221–2
    proposed by J. Locke, 219–20
    proposed by J. Swift, 214–18,
       222–3
    proposed by T. Hobbes,
       218–19
politics
    interpersonal relationships,
       207–8, 234–6
    and language, 233–4
    and literature, 62, 65–7, 80–8,
       328–34
    and writers, 311–12
polyandry, 205
polygamy, 205
pornography, 185
poststructuralism, 46–8, 102
postcolonial criticism, 44–5
power, 231–2, 267–8
    disciplinary, 261
    in family unit, 203–6
    Foucault's knowledge/power
       argument, 253–5
    and language, 232–3, 250–2
    misuse, 199
    resistance to, 265–6
    in society, 208
power relationships, 223–4, 337

and the Panopticon, 261–4
in society, 257
printing press, impact of, 247–8
prison system, 258–60
product/desire relationship,
    112–13
protest literature, 81–4
psychoanalytic criticism, 49
psychotherapy, 257, 264
'public utility', 227
Publications Act (1974), 328
Publications Appeal Board
    banning of *Dark Voices Ring*,
       328–34
punishment, 258–60
    as demonstration of sove-
       reign's power, 260
    penal reform, 260

Q

queens *see* monarchies
queer theory, 49

R

rape in poetry, 196–201
rationality, 220, 256
reader-response criticism, 25–6,
    42
reader's subjective reaction, 22,
    24
reality *versus* the 'real', 106, 113
reason *see* rationality
relationship of product/desire,
    112–13
religion, 39–40
resistance to power, 265–6
reviews, 88–9, 324
    Biting Through, by Sue
       Clark, 18–21, 30–1
    Dark Voices Ring, by Zakes
       Mda, 325–6
    Disgrace, by J.M. Coetzee,
       59–63
    These Things Happen, by

Shaun de Waal (reviewed
    by Tony Morphet), 89–92
revolution, 224–5, 229–30
rhetoric, 236, 241
'right to resist oppression', 228
'rights of the individual', 227
romantic love, 142
Rousseau, Jean Jacques, 221–2
run-on line, 11
Rushdie, Salmon *Satanic Verses*,
    72
Russian Formalists, 40
    see also formalist evaluation

S

Sachs, Albie *Preparing Ourselves
    for Freedom*, 84–8
Said, Edward, 44
San rock art interpretation, 29
Schreiner, Olive *Story of an
    African Farm*, 78–9
selling strategies, 121
semiotics, 102, 323
sensuality in love poetry, 145–7,
    151–2, 157–63, 178–80
Serote, Mongane Wally *For Don
    M. – Banned*, 17–18
setting analysis, 287–9
sex (physical characteristics), 118
sexuality, 95, 117, 137
    in advertising, 117–18
    in poetry, 184–95
sexually explicit language, 184–5
Shakespeare, William
    'Friends, Romans, countrymen
       ...' speech (from *Julius
       Caesar*), 294
    My Mistress' Eyes, 164–9
Shelley, Percy Bysshe *A Defence
    of Poetry*, 10
Shklovsky, Viktor, 10, 40
Showalter, Elaine, 45
Sidney, Sir Philip *Astrophel and
    Stella*, 154–5

simile, 11, 145–7
slogan, 110
Snailpress, 32
social control, 257–8
social conventions, 183–5
social history
    treatment of 'deviants'
       (Foucault), 254–7
social organizations, 206
Song of Solomon, (Bible) 143–8
sonnets, 165
South African literature
    'new' criteria, 84
speaker, 5
speakers in poetry, 145
speeches see political speeches
stanzas, 150
Sting *Sister Moon*, 169–72
Strong, Patience, 70–1
structuralism *see* poststructu-
    ralism
subjectivist interpretation, 25
Swift, Jonathan *Gulliver's Travels*,
    214–18, 222–3
syllabuses and 'political' power,
    250–2, 335
syllogism, 159
symbolism, 102, 327

T

'taboo', 184
target market, 109
text (definition), 101–2
textual analysis
    assumptions and implica-
       tions, 299–302, 319–20,
       326–7
    characterization, 289–92
    plot and theme, 292–9
    setting, 287–9
theatrical productions, 322–4
theme, 109, 178
    analysis, 292–9

Thiong'o, Ngũgĩ wa 'Literature
    and Society', 268–70
Tonkin, Boyd, 58
transformation (in S.A.), 84
'transgress', 183
translation, 148
Trapido, Barbara
    and *Disgrace*, by J.M. Coetzee,
       60

U

United States
    Declaration of Independence
       (1776), 226
    and Martin Luther King,
       236–8
university syllabuses and
    'political' power, 250–1, 335
utterance, 4, 6

W

Walsh, John
    and *Disgrace*, by J.M. Coetzee,
       61
women
    advertising stereotypes, 127
    fashions, 129–30
    image in poetry, 143–73
    and 'male gaze', 131–2
    notions of hysteria, 264
    power of decision making,
       234–5
World War I
    impact on literary studies, 39
writers and political views, 311
Wyatt, Michael *They flee from
    me*, 156–7